AN INTRODUCTION TO CATEGORY THEORY

AN INTRODUCTION TO
CATEGORY THEORY

V. SANKRITHI KRISHNAN
Professor of Mathematics
Temple University
Philadelphia

NORTH HOLLAND
New York • Oxford

Elsevier North Holland, Inc.
52 Vanderbilt Avenue, New York, New York 10017

Distributors outside the United States and Canada:

Elsevier/North-Holland Biomedical Press
335 Jan van Galenstraat, Box 211
Amsterdam, The Netherlands

Library of Congress Cataloging Data

Krishnan, Viakalathur Sankrithi, 1919–
An introduction to category theory.

Bibliography: p.
Includes index.
1. Categories (Mathematics) 2. Functor theory. I. Title.
QA169.K74 512'.55 80-14055
ISBN 0-444-00383-5

Desk Editor Louise Calabro Schreiber
Design Edmée Froment
Art Editor Aimee Kudlak
Art rendered by Vantage Art, Inc.
Production Manager Joanne Jay
Compositor Maryland Composition
Printer Haddon Craftsmen

Manufactured in the United States of America

To
Professor R. Vaidyanathaswamy

CONTENTS

PREFACE

Category theory is a relatively young branch of mathematics that has been growing quite rapidly both in content and applicability to other branches of the subject. As Professor MacLane's book title [22] suggests, all working mathematicians require some knowledge of it. So it was that I gave a course and ran seminars on category theory during the past few years at Temple University. This book is an outgrowth of those. There was, I felt, a need for a sufficiently short and coherent treatment of the basic notions and results for a first course or for a first reading by oneself. This *Introduction* would, I hope, provide such a treatment.

To keep the coverage sufficiently short, many topics of active study and interest had to be dropped; there is no mention in this book of homological algebra, Abelian categories, or topos theory. The text ends with a treatment of adjoint functors.

To provide the needed examples and background, a first chapter gives a survey of algebra and general topology. To make it easier for a first course, some of the longer proofs have been relegated to the end of the chapter under Exercises and Remarks. One deviation from normal: I treat the nonsymmetric forms in the semimetric and semiuniform spaces, since these have really more power in relating a preorder to the basic space (as is clear in the tract on topology and order by L. Nachbin[26]).

The next six chapters run through the basic notions and important results from category theory. Here again some special topics relating to feeble functors and weak adjoints are found at the ends of Chapters 4, 5, and 7, which are useful in dealing with types of extensions of ordered sets like the famous MacNeille extension of an ordered set to a complete lattice. But these extra topics could be skipped over at a first reading without losing continuity.

Similarly, as material suitable for a second reading, Appendixes 1, 2, and 3 treat of algebras over semiuniform spaces, over bitopological spaces, and over preordered sets, and of algebraic functors and topological functors. It is seen that a passage from a category of this type of object, which is part algebraic and part relational to another such, is an algebraic functor if the passage involves forgetting part of the algebraic structure, whereas it would be a topological functor if the passage involves forgetting part of the relational structure (Lemmas A.2.1 and A.3.1).

Finally, it is a pleasure to acknowledge the help I received from various persons in the preparation of this manuscript: to the students and colleagues who participated in my course and seminars; to my Ph.D. students, Malcolm Riddle and Leon Panichelli, for discussions I had with them on their work on duality and weak adjoints; to Dr. M. Riddle and to my wife, Dr. Marakatha Krishnan, for their careful weeding out of many errors in the manuscript; to the early reviewers of the manuscript whose remarks led to an improved and modified version; and to the Senior Desk Editor at Elsevier, Louise Schreiber, for her supervision of the careful copy editing of the manuscript. I would also like to thank the publishers, Elsevier North Holland, and in particular Senior Editor Kenneth J. Bowman for the gracious and efficient way they transformed the manuscript to the printed book.

AN INTRODUCTION TO
CATEGORY THEORY

CHAPTER ONE

BASICS FROM ALGEBRA AND TOPOLOGY

1.1 SET THEORY

We take as "primitive notions" the *set* and the relation of *membership* of an element in a set. We write $x \in A$ when x is a member of the set A; sometimes the same relationship is written as $A \ni x$. In terms of this relationship we can define two others for sets: $A \subseteq B$ (read A is contained in B) and $B \supseteq A$ (read B contains A), both of which are taken to mean that $[x \in A]$ implies $[x \in B]$ for all x. And when each of A, B is contained in the other (and only then), we write $A = B$. We admit a set that contains no elements; it is called a *null set* and is usually denoted by the symbol $\varnothing$.

Sets A, B determine a set $A \times B$, called the product of the sets A and B, which consists of elements of the form (a, b), where $a \in A$, $b \in B$; these elements of $A \times B$ are the "ordered pairs," a first term from A and a second term from B. For $A \times A$ we also write A^2.

By a relation from the set A to the set B we mean any subset of $A \times B$; we also say that ρ is a relation on A if it is a subset of $A \times A$. Given a relation ρ from A to B, and a relation σ from B to C, for any sets A, B, and C we now define

1. the relation ρ^r from B to A by $(b, a) \in \rho^r$ iff $(a, b) \in \rho$;
2. the relation $\sigma \circ \rho$ from A to C by $(a, c) \in (\sigma \circ \rho)$ iff, for some b in B, $(a, b) \in \rho$ and $(b, c) \in \sigma$;
3. for a subset C of A, $\rho(C) = [b \in B$: for some c of C, $(c, b) \in \rho]$; and
4. for an element c of A, $\rho(c) = \rho(\{c\})$, where $\{c\}$ is a one-element set.

The relation ρ^r is called the *reverse of the relation* ρ; and $(\sigma \circ \rho)$ is called the *relational product* of ρ followed by σ. For the relation ρ from A to B, we set $\text{dom}(\rho) = [c \text{ in } A\colon \rho(c) \text{ is nonnull}]$ and call it the domain of ρ, and we set $\text{ran}(\rho) = [b \text{ in } B\colon \rho^{\text{r}}(b) \text{ is nonnull}]$ and call it the range of ρ.

Among the relations on a set A is $1_A = [(a, a)\colon a \text{ in } A]$; this is called the *identity relation* on A. For any relation ρ on A we say that ρ is

1. a *reflexive relation* if ρ contains 1_A;
2. a *symmetric relation* if ρ contains (and hence equals) ρ^{r};
3. a *transitive relation* if ρ contains $(\rho \circ \rho)$;
4. an *equivalence relation* if it is reflexive, symmetric, and transitive;
5. a *preorder relation* if it is reflexive and transitive;
6. an *order relation* if it is a preorder relation and further (a, a') belongs to ρ and to ρ^{r} only when $a = a'$.

Next, for a relation ρ from A to B we say that ρ is

1. *injective* if, for each b in B, $\rho^{\text{r}}(b)$ has at most one element;
2. *surjective* if the range of ρ is B;
3. *bijective* if it is both injective and surjective;
4. a *partial mapping* if ρ^{r} is injective;
5. a *cofull relation* if ρ^{r} is surjective;
6. a *mapping relation* (or just a *mapping*) if ρ^{r} is bijective;
7. *bi-injective* if both ρ and ρ^{r} are injective.

It is clear that when ρ is a mapping relation, each element c of A determines a unique element b of B such that $\rho(c) = (b)$; we usually write then $\rho(c) = b$ and call b the image of c under the mapping ρ. If ρ is bijective and a mapping, hence a bijective mapping, it really gives rise to a one-to-one correspondence $[c\text{–}b]$ between the elements of A and B such that $\rho(c) = b$ and $\rho^{\text{r}}(b) = c$. It is not hard to see that ρ is bijective if and only if $\rho^{\text{r}} \circ \rho = 1_A$ and $\rho \circ \rho^{\text{r}} = 1_B$. Finally, a bi-injective relation ρ from A to B may also be viewed as a bijective mapping of $\text{dom}(\rho)$ on $\text{ran}(\rho)$.

The relation $\subseteq$ of set-inclusion is always an order relation on any family of sets $\mathbf{F}$. For any subfamily $\mathbf{A}$ of $\mathbf{F}$, we have a *union* $\cup(\mathbf{A})$ and an *intersection* $\cap(\mathbf{A})$, defined by $\cup(\mathbf{A}) = [x\colon x \text{ is in one of the sets of } \mathbf{A}]$ and $\cap(\mathbf{A}) = [x\colon x \text{ is in each set of } \mathbf{A}]$. We see that $\cup(\mathbf{A})$ is the smallest set containing each set of $\mathbf{A}$ and $\cap(\mathbf{A})$ is the largest set contained in each set of $\mathbf{A}$. For a pair of sets A, B we write $A \cup B$, $A \cap B$ for their union and intersection, respectively. With this notation we can verify also that for any family of sets $\mathbf{A}$ and a single set B we have the *distributive laws* $B \cap [\cup(\mathbf{A})] = \cup[(B \cap A_i)\colon A_i \text{ ranging over } \mathbf{A}]$ and $B \cup [\cap(\mathbf{A})] = \cap[(B \cup A_i)\colon A_i \text{ in } \mathbf{A}]$.

The family $P(X)$ of all subsets of a given set X is called the *power set* over X; it is closed for formation of unions and intersections of all subfamilies. Further, any A of $P(X)$ determines a unique *complement* $C(A)$ in $P(X)$ given by $C(X) = [x$ in X: x is not in $A]$.

A family of sets $[X_i$: i in $I]$ indexed by a set I determines a *direct product* set $P(X_i)$ and a *disjoint union* set $\Sigma^\circ(X_i)$; these are given by $P(X_i) =$ [mapping $\alpha: I \to \cup(X_i)$: $\alpha(i)$ is in X_i for each i] and $\Sigma^\circ(X_i) = [(i, x)$: i is from I and x from $X_i]$. We assume the "axiom of choice," which assures that $P(X_i)$ is nonnull when I and the X_i are nonnull. From the direct product set $P(X_i)$ there are *canonical mappings* p_i: $P(X_i) \to X_i$ to the X_i given by $p_i(\alpha) = \alpha(i)$ for any α in $P(X_i)$. Further, given a set C and a family of mappings g_i: $C \to X_i$, one for each i of I, a unique map $g = SP(g_i)$, called the *semiproduct* of the family (g_i), can be defined in such a way that $g_i = p_i \circ g$ for each i, namely, by setting $g(c) = \alpha$, where $\alpha(i) = g_i(c)$ for each i. Similarly, there are maps j_i: $X_i \to \Sigma^\circ[X_i]$, one for each i, given by $j_i(x) = (i, x)$ for x in X_i; and for a given set D and a family of maps h_i: $X_i \to D$, one for each i, a unique map h can be defined from $\Sigma^\circ[X_i]$ to D such that $h_i = h \circ j_i$ for each i by setting $h(i, x) = h_i(x)$ for any (i, x) from $\Sigma^\circ[X_i]$.

For a finite family $[X_1, \ldots, X_n]$ of sets the direct product is also denoted by $X_1 \times \cdots \times X_n$ and the disjoint union by $X_1 \mathring{\cup} \cdots \mathring{\cup} X_n$.

1.2 SOME TYPICAL ALGEBRAIC STRUCTURES

A set X is said to be *closed for the n-ary operation* θ if to each ordered n-tuple of elements $(x_1, \ldots, x_n)$ from X a unique element $\theta(x_1, \ldots, x_n)$ of X is associated. An n-ary operation for $n = 1, 2$, or 3 is referred to as a *unary, binary,* or *ternary* operation, respectively. We also talk of a *nullary* operation, corresponding to $n = 0$: such an operation associates to the null set $\varnothing(\subseteq X)$ a unique element $\theta(\varnothing)$. Thus it is an operation that chooses a distinguished element from X.

For a binary operation θ on X it is usual to write $(x \,\theta\, y)$ for $\theta(x, y)$. Often the symbol used for this operation is just a dot $(\cdot)$, so that $x \cdot y$ stands for the effect of the operation on (x, y). We call such an operator a *multiplication,* and $x \cdot y$ is called the product of (x, y).

The binary operation θ on X is said to be (a) *commutative* if $x \,\theta\, y = y \,\theta\, x$ for all x, y from X; (b) *associative* if $x \,\theta\, (y \,\theta\, z) = (x \,\theta\, y) \,\theta\, z$, for all x, y, and z from X; (c) *idempotent* if $x \,\theta\, x = x$ for all x of X.

In a set X closed for the binary operation θ, an element c of X is said to be (a) an *idempotent element* if $c \,\theta\, c = c$; (b) a *left unit element* if $c \,\theta\, x = x$ for each x of X; (c) a *left zero element* of X if $c \,\theta\, x = c$ for each x in X; (b') a *right unit element* if $x \,\theta\, c = x$ for each x; and (c') a *right zero element* if $x \,\theta\, c = c$ for all x in X. When c is both a left and a right unit (or both a left and a right zero) element in X, it is called just a *unit* (or *zero*) *element* of X. When we wish to specify the operation too, we say that c is a unit (or left zero) element of (X, θ).

A *groupoid* (X, θ) is a set X closed for a binary operation θ. A groupoid (X, θ) is called a *semigroup* if θ is associative. A semigroup (X, θ) is called a *semilattice* if θ is commutative and idempotent.

For any ordered set of n elements $(x_1, \ldots, x_n)$ of a groupoid (X, θ) we can define inductively a product element denoted by $x_1 \theta x_2 \theta \cdots \theta x_n$ by setting $x_1 \theta \cdots \theta x_{n+1}$ to be $[x_1 \theta x_2 \theta \cdots \theta x_n] \theta x_{n+1}$ for $n \geq 2$. We also have a generalized associative law in a semigroup (X, θ); for integers m, $n(>0)$ and any $m + n$ elements $x_1, \ldots, x_{m+n}$ from X we have the equality $(x_1 \theta \cdots \theta x_m) \theta (x_{m+1} \theta \cdots \theta x_{m+n}) = x_1 \theta \cdots \theta x_{m+n}$. This can be proved inductively.

We say that y is a *relative inverse* of an element x in a semigroup (X, θ) if $x \theta y \theta x = x$ and $y \theta x \theta y = y$; when each element of (X, θ) has (at least) a relative inverse we call the semigroup a *regular semigroup*. By choosing one such relative inverse for each x and denoting it by $(x)'$, we can think of a regular semigroup as an algebra $(X, \theta, (\ )')$ with an extra unary operation $(\ )'$ such that $x \theta (x)' \theta x = x$ and $(x)' \theta x \theta (x)' = (x)'$.

When $e^2 = e \theta e$ is the same as e for an element e of the semigroup (X, θ), we call e an *idempotent element* of the semigroup. When $(X, \theta, (\ )')$ is a regular semigroup in which each pair of idempotent elements e, f satisfies the commutativity condition $e \theta f = f \theta e$, the regular semigroup is called an *inverse semigroup*. We shall see later that in such an inverse semigroup each x has only one relative inverse.

A *group* can be defined as an inverse semigroup in which there is only one idempotent.

A mapping f: $X \to X^*$ is called a *groupoid homomorphism*/[a *regular groupoid homomorphism*] of the groupoid (X, θ) in (X^*, θ^*)/[of the regular semigroup $(X, \theta, (\ )')$ in the regular semigroup $(X^*, \theta^*, (\ )'^*)$] if for any x, y from X, $f(x \theta y) = (f(x)) \theta^* (f(y))$/[and further $f((x)') = (f(x))'^*$]. If in the foregoing X is a subset of X^* and f is the inclusion map of X in X^*, and f is such a homomorphism, we call (X, θ) a *subgroupoid* of (X^*, θ^*)/[call $(X, \theta, (\ )')$ a *subregular groupoid* of $(X^*, \theta^*, (\ )'^*)$]. It can be seen that a subgroupoid of a semigroup is a semigroup and that a subregular groupoid of an inverse semigroup or a group is also an inverse semigroup or group.

Theorem 1.2.1 *(a) A group $(X, \cdot, (\ )', n)$ can be characterized as a set closed for three operations—a binary $(\cdot)$, a unary $(\ )'$, and a nullary (n)—such that for all x, y, z from X, (i) $x \cdot (y \cdot z) = (x \cdot y) \cdot z$, (ii) setting $n(\varnothing) = 1$, we have $1 \cdot x = x = x \cdot 1$, and (iii) $x \cdot x' = 1 = x' \cdot x$ (setting x' for $(x)'$).*

(b) A group is just an inverse semigroup with only one idempotent element, whereas a semilattice is just an inverse semigroup all of whose elements are idempotent elements.

PROOF. For (a) the proof follows from the fact that in a regular semigroup $x \cdot x'$ and $x' \cdot x$ are idempotent elements for any x. The proof of (b) follows if we note that a semilattice is always an inverse semigroup. □

When a groupoid homomorphism or regular groupoid homomorphism is a bijective map, it is easy to see that the reverse map is also such a homomorphism; we call the homomorphism in this case *an isomorphism* between the groupoids or the regular groupoids. Such isomorphic groupoids have essentially the same properties and are usually considered "equal."

We now describe a class of semigroups that provide isomorphs of all semigroups of special types. The family $R(A)$ of all relations on a set A (taken to be nonnull) forms a semigroup under the relational multiplication operation ($\circ$). There are subgroupoids of this semigroup $(R(A), \circ)$ formed by taking the following subsets of $R(A)$ as elements: (i) the set $PM(A)$ of all partial mappings; (ii) the set $M(A)$ of all mappings; (iii) the set $BI(A)$ of all bi-injective relations; and (iv) the set $BM(A)$ of all bijective mappings. Of these, $BM(A)$ is contained in $BI(A)$ and $M(A)$, while these two are both contained in $PM(A)$.

Theorem 1.2.2 *(a) For each element of $PM(A)$ there exist a number of relative inverses, so that $(PM(A), \circ)$ is a regular semigroup with unit element. Similarly, $(M(A), \circ, (\ )')$ is also a regular semigroup with unit element with many possible choices of relative inverses for elements.*

(b) $(BI(A), \circ, (\ )')$ is an inverse semigroup with a unique choice of a relative inverse for each element, and $(BM(A), \circ, (\ )')$ is a group.

(c) Any semigroup (X, θ) with unit element is isomorphic with a subgroupoid of $(M(X), \circ)$. Any regular semigroup $(X, \theta, (\ )^*)$ with unit element is isomorphic with a subregular groupoid of $(M(X), \circ, (\ )')$ for a suitable choice of relative inverse for each element. Any inverse semigroup $(X, \theta, (\ )^*)$ is isomorphic with a subregular groupoid of $(BI(X), \circ, (\ )')$; and any group $(X, \theta, (\ )^*)$ is isomorphic with a subregular groupoid of $(BM(X), \circ, (\ )')$.

For the proof of the theorem see Exercises and Remarks at the end of this chapter, page 23.

1.3 ALGEBRAS IN GENERAL

A pair (X, F) is called an *algebra of type F* (or an *F-algebra*) if X is a set, F is a family of pairs $[(\theta_i, n_i)$: i in $I]$, each θ_i being the symbol for an abstract operation with "arity" n_i (n_i an integer ≥ 0), and there is defined a mapping $\theta_i^X: (X)^{n_i} \rightarrow X$, for each pair (θ_i, n_i) from F. The element $\theta_i^X[(x_1, \ldots, x_{n_i})]$ in X, associated to a given n_i-tuple of elements $(x_1, \ldots, x_{n_i})$ from X, is called the result of the operation θ_i on the n_i-tuple. When $n_i = 0$, we assume that $\theta_i(\varnothing)$ equals a uniquely determined element from X; θ_i is now defined on the null set only. Thus these nullary operations essentially choose special elements from X. For $n = 1, 2, 3$, etc., an operation of arity n is called unary, binary, ternary, etc. We already met with the unary, binary, and nullary operations in defining a group. The group can be thought of as an F-algebra for $F = [(\theta, 2), ((\ )', 1), (e, 0)]$. Thus $e(\varnothing)$ = the unit element of the group.

Given two F-algebras (X, F) and (Y, F), a mapping f: $X \to Y$ is called an *F-homomorphism* from (X, F) to (Y, F) if for each (θ_i, n_i) of F and each n_i-tuple $(x_1, \ldots, x_{n_i})$ from X we have $f(\theta_i^X(x_1, \ldots, x_{n_i})) = \theta_i^Y[f(x_1), \ldots, f(x_{n_i})]$. If it also happens that the mapping f: $X \to Y$ is bijective, then it will be seen that when f is a homomorphism from (X, F) to (Y, F), the reverse map f^r is also a homomorphism from (Y, F) to (X, F); we then call f an *isomorphism* from (X, F) to (Y, F) and, similarly, f^r an isomorphism from (Y, F) to (X, F). These isomorphic algebras have essentially the same structural properties.

Given an F-algebra (X, F) and a subset Y of X, we say that Y is closed for one of the operations (θ, n) from F if for any n-tuple of elements $(y_1, \ldots, y_n)$ from Y, $\theta^X(y_1, \ldots, y_n)$ is an element of Y. If the subset Y is closed for each operation from F, then we get an F-algebra (Y, F), called an *F-subalgebra* of (X, F) when we set $\theta^Y(y_1, \ldots, y_n)$ to be the same as $\theta^X(y_1, \ldots, y_n)$ for each n-tuple of elements $(y_1, \ldots, y_n)$ from Y and each (θ, n) from F. For example, our earlier notions of subgroupoid and subregular groupoid are special cases of F-subalgebras. Evidently the inclusion mapping of Y in X is an F-homomorphism of (Y, F) in (X, F) when (Y, F) is an F-subalgebra of (X, F).

Given any indexed family of F-algebras $[(X_i, F)$: i in $I]$, we form an F-algebra $(X, F) = P[(X_i, F)$: i in $I]$, called the *product* of the given family of F-algebras, by taking X to be the product set $P[X_i$: i in $I]$ and, for a (θ, n) of F and an n-tuple of elements $(x_1, \ldots, x_n)$ from X, setting $\theta^X(x_1, \ldots, x_n) = x$ in X such that $p_i(x)$ = the ith component of $x = \theta^{X_i}[p_i(x_1), \ldots, p_i(x_n)]$ for each i from I. It is clear that the canonical projection maps p_i: $X \to X_i$ are F-homomorphisms of (X, F) in the (X_i, F).

An equivalence relation E on the set X is called a *congruence relative to* an operation (θ, n) from F *on* the F-algebra (X, F) if, given an n-tuple of pairs $[(x_i, y_i)$: $i = 1, \ldots, n]$, all from E, the pair $[\theta^X(x_1, \ldots, x_n), \theta^X(y_1, \ldots, y_n)]$ also belongs to E. If E is a congruence relative to each (θ, n) from F, we call it just a *congruence on the F-algebra* (X, F). In this case, we can define a *quotient F-algebra* $(X, F)/E = (X/E, F)$, with the quotient set X/E as the set, by writing the result of θ for a (θ, n) of F and an n-tuple of elements $[x_1^E, \ldots, x_n^E]$ of X/E to be $\theta^{(X,F)/E}[x_1^E, \ldots, x_n^E] = [\theta^X(x_1, \ldots, x_n)]^E$, where we set x^E to denote the E-class containing the element x of X. That the result of the operation so defined is really independent of the choice of the elements x_i from the classes containing them follows from our definition of a congruence; and the canonical surjective mapping of X on X/E is now a homomorphism of (X, F) on $(X/E, F)$.

Lemma 1.3.1 *(a) A homomorphism $f: (X, F) \to (Y, F)$ determines a congruence E on (X, F) and also a subalgebra $(f(X), F)$ of (Y, F) such that $f = j \cdot i \cdot p$ where p is the canonical surjection of (X, F) on $(X/E, F)$, i is an isomorphism of $(X/E, F)$ on $(f(X), F)$, and j is an inclusion map and homomorphism of the subalgebra $(f(X), F)$ in (Y, F).*

(b) The family of congruences on an algebra (X, F) forms a complete lattice when ordered by the relation $\subseteq$ of set-inclusion; the lattice products in this lattice are the intersections of congruence families.

PROOF. (a) Set $E = [(x, x'): f(x) = f(x')]$; note that this is a congruence on (X, F) and that the subset $f(X)$ of Y is closed under F, so it gives a subalgebra $(f(X), F)$ of (Y, F).

(b) Check that the set-intersection of any nonnull family of congruences is a congruence, and that there is a largest congruence, namely, $X \times X$. □

An F-algebra (X, F) is called an *irreducible F-algebra* if [the intersection of a family of congruences on (X, F) equals 1_X] implies that [one of the congruences of the family equals 1_X]. With this definition, we have a fundamental theorem due to Birkoff ([2], Chapter VIII, Section 8, Theorem 15):

Theorem 1.3.1 *Given any F-algebra (X, F), we can find a family $[C_j: j$ in $J]$ of congruences on (X, F) such that (i) each of the quotient F-algebras $(X/C_j, F)$ is an irreducible F-algebra; (ii) (X, F) is isomorphic with a subalgebra (X', F) of the product $P[(X/C_j, F): j$ in $J]$; and (iii) this subalgebra (X', F) is a large subalgebra of the product algebra, meaning thereby that for each j of J and each element x^{C_j} of (X/C_j) there is an element x' of X' such that $x'(j) = x^{C_j}$.*

PROOF. Let J denote the set $[(x, y) \in X \times X: x \neq y] = (X \times X) - 1_X$. For each (x, y) from J we choose a maximal congruence $C(x, y)$ on (X, F) not containing the pair (x, y): such maximals exist, by Zorn's principle, since there is one such congruence (1_X) and the union of a chain of such congruences (under the order $\subseteq$) is also such a congruence and an upper bound for the elements of the chain. If we define the map f: $(X, F) \to P[(X, F)/C(x, y): (x, y) \in J]$ by $f(z) = z'$ such that $z'(x, y) =$ the class under $C(x, y)$ containing z, then f is a homomorphism of F-algebras and is an injective map, too; for if $x \neq y$ for x, y from X, then $f(x) \neq f(y)$, since there is a j $(=(x, y))$ such that $C(j)(x) \neq C(j)(y)$: the $C(j)$-classes containing x and y are distinct. Thus f is an isomorphism of the original F-algebra with a subalgebra (X', F) of the product algebra. This image algebra is large in the product, since any element $C(j)(z)$ from $X/C(j)$ is $f(z)(j)$ for a certain z from X. □

Although a groupoid can be identified just as an F-algebra with $F = (\theta, 2)$, a semigroup has to satisfy an identical equation of the form $(x \ \theta \ y) \ \theta \ z = x \ \theta \ (y \ \theta \ z)$ for all x, y, z from the groupoid. To see how to deal suitably with such identical relations, we start with constructing a free F-algebra over any set A. Thus, we begin with a nonnull set A and a nonnull family $F = [(\theta_j, n_j): j \text{ in } J]$ of abstract operations, each with an arity (an integer $n_j \geq 0$); the symbols θ_j thus are just names of operations. The *free F-algebra over A* is of the form $[P(A, F), F]$, where the set $P(A, F)$ consists of what we call *polynomial expressions in F over A*, each of them with a uniquely associated integer ≥ 0, called its *degree*. These we define inductively as follows. A polynomial of degree 0 is either an element of A or one of the symbols for a nullary operation from F; that is, it is of the form [b, for a b in A; or θ, for a $(\theta, 0)$ of F]. When the polynomials of degree $\leq n$ have been defined, for some integer $n \geq 0$ we define a polynomial of degree $(n + 1)$ to be of the form $\theta[x_1, \ldots, x_m]$ where (θ, m) is an element of F with $m \geq 1$, the elements $x_1, \ldots, x_m$ are polynomials of degrees $\leq n$, and at least one of them is of degree n. Here the expression $\theta[x_1, \ldots, x_m]$ is just a formal one involving the symbol θ and the symbols of earlier defined polynomials. For this collection $P(A, F)$ of all such polynomials (each of some finite degree) we can define the operations from F: for a (θ, n) of F with $n = 0$, $\theta^{P(A,F)}(\varnothing) = \theta$ considered as a polynomial of degree 0; for a (θ, n) of F with $n \geq 1$ and a collection of n polynomial symbols $(x_1, \ldots, x_n)$ from $P(A, F)$, $\theta^{P(A,F)}(x_1, \ldots, x_n) =$ the polynomial expression $\theta[x_1, \ldots, x_n]$, which belongs to $P(A, X)$ and is of degree $(m + 1)$, if $m =$ the maximum of the degrees of the polynomials $\{x_1, \ldots, x_n\}$. It is not hard to see that this construction indeed gives an F-algebra, and that the set base $P(A, F)$ of this algebra contains the set A as a subset. More is true:

Lemma 1.3.2 *There is an injective map, namely, the inclusion map* $1_{A \subset P(A,F)}$ *of the set A in the set base P(A, F) of the free F-algebra* $[P(A, F), F]$, *and if there is a set map* $g: A \to Y$ *of A in the set base of any F-algebra (Y, F), then there is a unique F-homomorphism* g^*: $[P(A, F), F] \to (Y, F)$ *such that* $g^* \cdot 1_{A \subset P(A,F)} = g$. *This unique* g^* *is called the extension of the map g to a homomorphism from the free algebra over A to (Y, F).*

PROOF. We define g^* inductively for the elements of $P(A, F)$. For an element p of degree 0 from $P(A, F)$, we set $g^*(p) = \theta^Y(\varnothing)$ if p is of the form θ for a $(\theta, 0)$ from F, and we set $g^*(p) = g(p)$ if p is an element of A. When g* has been defined for all elements of degree $\leq n$ (for $n \geq 0$), for an element $p = \theta[x_1, \ldots, x_m]$ of degree $n + 1$ from $P(A, F)$ we set $g(p) = \theta^Y(g^*(x_1), \ldots, g^*(x_m))$ and note that all these elements $g^*(x_i)$ are now well-defined elements of Y, since the x_i have degree $\leq n$. This shows that $g^* \cdot 1_{A \subset P(A,F)} = g$; and it is not hard to see that g^* is an F-homomorphism of $[P(A, F), F]$ in (Y, F), while if there were any homomorphism g'' of $[P(A, F), F]$ in (Y, F) such that $g'' \cdot 1_{A \subset P(A,F)} = g$, it could be shown by inductive arguments that g'' coincides with g^* for elements of degree 0, degree 1, and so on for all elements of $P(A, F)$. This gives the uniqueness of g^* as asserted. □

Corollary *Any F-algebra (X, F) is isomorphic with the quotient algebra by a suitable congruence of the free F-algebra [P(X, F), F] over the set X. (Because there is an injective map x → x of the set X in the set base X of the F-algebra (X, F), and the extension of this to a homomorphism from [P(X, F), F] to (X, F) is surely surjective.)*

Now we are in a position to define what are called *identical relations* in an F-algebra. Assume that we have a free F-algebra $[P(A, F), F]$ over a certain nonnull set A. Let D be a family of pairs of elements $[(p_i, q_i)$: i in $I]$ from $P(A, F)$. Then we say that an *F-algebra* (X, F) *satisfies the identical relations* $[p_i = q_i$: i in $I]$ if for each set map $g: A \to X$ the extension map $g^*: P(A, F) \to X$ is such that $g^*(p_i) = g^*(q_i)$ in X for each of the pairs (p_i, q_i) from D. When it satisfies these relations we call it an *FD-algebra*.

Lemma 1.3.3 *Given the family F of operations and a family D of pairs from P(A, F) and a set X, there are congruences (E_j) on the free F-algebra (P(X, F), F) that contain all pairs from P(X, F) of the form [($g^*(p_i)$, $g^*(q_i)$): for all (p_i, q_i) from D and for all maps g: A → X and the associated* g*: *P(A, F) → P(X, F)]; the intersection of all such congruences is the least congruence in the family. If this is denoted by E*, the quotient-algebra (P(X, F)/E*, F) is an FD-algebra. This we call a free FD-algebra over X. There is a canonical set mapping k: X → (P(X, F), F) such that for any set map f: X → Y of X when (Y, F) is an FD-algebra, there is a unique F-homomorphism f# of (P(X, F)/E*, F) in (Y, F) such that f = f# · k.*

PROOF. Denoting by j the canonical injective map of X in $P(X, F)$, we know (from Lemma 1.3.2) that the map f: $X \to Y$ can be resolved in the form $f = f^* \cdot j$, where f^* is an F-homomorphism of $(P(X, F), F)$ in (Y, F). If p denotes the canonical surjective F-homomorphism of $(P(X, F), F)$ on $(P(X, F)/E^*, F)$, we define $k = p \cdot j$. From the fact that E^* contains all pairs of the form $[(g^*(p_i), g^*(q_i)]$ it follows that the quotient algebra must be an *FD*-algebra. Further, since (Y, F) is assumed to be an *FD*-algebra, the images under f^* of the pair $g^*(p_i)$, $g^*(q_i)$ must be equal, because $f^*g^*(p_i) = (f \cdot g)^*(p_i)$ and $f^*g^*(q_i) = (f \cdot g)^*(q_i)$ for a mapping $f \cdot g$: $A \to Y$. This means that the congruence $f^{*r} \cdot f^*$ determined by f^* on $(P(X, F), F)$ contains all the pairs $[(g^*(p_i), g^*(q_i)]$, hence must be one of the E_j and must contain E^*. Then surely the morphism f^* splits as $f^* = f\# \cdot p$ for a suitable homomorphism $f\#$, and we have then $f = f^* \cdot j = f\# \cdot p \cdot j = f\# \cdot k$. □

The next lemma lists some further easily proved properties of *FD*-algebras.

Lemma 1.3.4 *(i) An F-subalgebra of an FD-algebra is an FD-algebra; (ii) the quotient algebra of an FD-algebra by an F-congruence is an FD-algebra; and (iii) a direct product of a set of FD-algebras is an FD-algebra.*

Using this lemma and Theorem 1.3.1, we get the following corollary:

Corollary *Any FD-algebra is isomorphic with a large F-subalgebra of a direct product of a family of FD-algebras that are all irreducible F-algebras.*

Given an *F*-algebra (Y, F), a subset X of Y is called a *generating set* for the *F*-algebra if the homomorphism $P(f)$: $(P(X, F), F) \to (Y, F)$ defined by the inclusion map f: $X \to Y$ is a surjective map. In that case (Y, F) would be isomorphic with a quotient algebra of $(P(X, F), F)$ by a congruence, and further, each element of Y could be expressed in terms of elements of X using the operations from F.

For example, a semigroup can be considered as an *FD*-algebra; and for the semigroup $(N, +)$ of the natural numbers under addition there is a generating set [1] with just one element 1.

1.4 TOPOLOGICAL SPACES

A topological structure for a set X (or, as we often call it, for a *space X* of *points*) can be assigned using one of many different auxiliary notions: by neighborhoods for points, by convergence of directed sequences of points, by a closure operator, by a family of closed sets, or by a family of open sets. The last is usually taken as the characteristic form. So we set the following:

Definition 1.4.1 A set X together with a family $\mathbf{G}$ of subsets of X, called the *open sets* of X, forms a *topological space* $(X, \mathbf{G})$ if the family $\mathbf{G}$

satisfies the following conditions: (01) the set X and the set $\varnothing$ are in **G**; (02) if A, B are sets from **G**, then $A \cap B$ is also in **G**; and (03) if $[A_i$: i in $I]$ is a family of sets from **G**, then $\cup(A_i$: i in $I)$ is also in **G**.

Definition 1.4.2 A *neighborhood space* (X, N) consists of a set X together with a mapping N of X in the power set $P[P(X)]$ such that (N1) for each x of X and for each A in $N(x)$, x is in A; these sets of $N(x)$ are called the $(N-)$ neighborhoods of x; (N2) if A, B are from $N(x)$ for some x of X, then $A \cap B$ contains some C of $N(x)$; (N3) if x is in X and A is in $N(x)$, there is a B in $N(x)$ such that A contains some C from $N(y)$ for each y in B.

Definition 1.4.3 For a down-directed set $(D, \leq)$ a mapping **s**: $D \rightarrow X$ is called a *$(D, \leq)$-sequence in X*; if $(E, \leq)$ is a down-directed set and k: $(E, \leq) \rightarrow (D, \leq)$ is an isotone map mapping E on a coinitial subset of $(D, \leq)$, then a $(D, \leq)$-sequence s in X determines an $(E, \leq)$-sequence $(s \cdot k)$, which is called a *subsequence of s*.

Definition 1.4.4 A *convergence space* (X, C) consists of a set X together with a family C of *convergences*: each convergence is a triple $[(D, \leq)$, $s, x]$ consisting of a down-directed set $(D, \leq)$, a $(D, \leq)$-sequence s (or map s: $D \rightarrow X$) in X, and a point x of X called a *limit point of the sequence s in X*, subject to the following conditions: (C1) for each down-directed set $(D, \leq)$ and any point x of X, if x^*: $D \rightarrow X$ is the constant function with $x^*(d) = x$ for each d of D, then $[(D, \leq), x^*, x]$ belongs to C; (C2) if k is an isotone map of the down-directed set $(E, \leq)$ on a coinitial subset of $(D, \leq)$ and if $[(D, \leq), s, x]$ belongs to C, then $[(E, \leq), s \cdot k, x]$ also belongs to C; and (C3) when $\{[(D, \leq), s, x] \in C$, and for each d of D an $[(E_d, \leq), s_d, s(d)]$ also belongs to $C\}$, then there is an $[(F, \leq), t, x]$ in C with the property that $t(F)$ is contained in $\cup[s_d(E_d)$: d in $D]$.

Definition 1.4.5 A *closure operator* Cl ( ) *on a set X* is a mapping $A \rightarrow \mathrm{Cl}(A)$ of $P(X)$ in itself such that (Cl.1) $A \subseteq B$ implies $\mathrm{Cl}(A) \subseteq \mathrm{Cl}(B)$; (Cl.2) for each A of $P(X)$, $A \subseteq \mathrm{Cl}(A)$, and $\mathrm{Cl}(\varnothing) = (\varnothing)$; (Cl.3) for all A, B from $P(X)$, $\mathrm{Cl}(A \cup B) = \mathrm{Cl}(A) \cup \mathrm{Cl}(B)$. A *closure space* $(X, \mathrm{Cl}(\ ))$ is a space X (of points) together with a closure operator Cl( ) on X.

Definition 1.4.6 By a *family of closed sets for X* is meant a family **F** of subsets of X satisfying the conditions: (F1) X and $\varnothing$ are in **F**; (F2) when A, B are in **F**, $(A \cup B)$ is also in **F**; and (F3) if $[A_i$: i in $I]$ is a family of sets from **F**, then $A = \cap[A_i$: i in $I]$ is also in **F**.

Definition 1.4.7 A nonnull family of nonnull subsets, $\mathbf{B} = [B_j]$, of a set X is called a *filter base over X* if $\mathbf{B}$ is down-directed under the order $\subseteq$: that is, each pair of sets B_i, B_j from $\mathbf{B}$ contains a set B_k from $\mathbf{B}$. Such a filter base $\mathbf{B}$ is called a *filter over X* if it satisfies the further condition [B_j in $\mathbf{B}$, and the subset C of X contains B_j] imply [C is in $\mathbf{B}$].

Definition 1.4.8 A filter base $\mathbf{B}$ over X determines a smallest filter containing it, denoted by $F(\mathbf{B})$; it is the family of subsets of X that contain sets from $\mathbf{B}$. A filter base $\mathbf{B}$ is said to be *finer than* another $\mathbf{B}'$ (and $\mathbf{B}'$ is said to be *coarser than* $\mathbf{B}$) if $F(\mathbf{B})$ contains $F(\mathbf{B}')$, or if each set of $\mathbf{B}'$ contains some set from $\mathbf{B}$. When $F(\mathbf{B})$ equals $F(\mathbf{B}')$ we call the filter bases $\mathbf{B}$ and $\mathbf{B}'$ *equivalent*. Clearly this relation of being "finer than" is just set-inclusion when applied to filters. A filter $\mathbf{F}$ is called an *ultrafilter* if [$\mathbf{G}$ is a filter finer than $\mathbf{F}$] implies that [$\mathbf{G} = \mathbf{F}$].

The following can be seen to be alternative conditions to define the ultrafilters among filters:

1. the filter $\mathbf{F}$ is an ultrafilter iff (the intersection $B \cap C$ of two subsets of X belongs to $\mathbf{F}$ only when one of B, C belongs to $\mathbf{F}$);
2. the filter $\mathbf{F}$ is an ultrafilter iff for each subset A of X either A or its complement A' belongs to $\mathbf{F}$.

When (X, N) is a neighborhood space, for each x of X, $N(x)$ is clearly a filter base over X; it is called the *N-base of neighborhoods at x*. The filter $F(N(x))$ it generates is called the *filter of neighborhoods at x* defined by N.

After these preliminaries, we relate the various forms of space structures.

Lemma 1.4.1 *(a) If (X, N) is a neighborhood space, it defines an associated convergence space (X, C) when we set $[(D \leq), s, x] \in C$ iff for each U in $N(x)$ there is a nonnull initial subset B of $(D, \leq)$ such that $s(b)$ is in U for each b from B.*

(b) Given a convergence space (X, C), it defines an associated closure space $(X, \mathrm{Cl}(\ ))$ when we set $\mathrm{Cl}(A) = [x$ in X: there is a $[(D, \leq), s, x]$ in C with $s(d)$ in A for each d of $D]$.

(c) Given a closure space $(X, \mathrm{Cl}(\ ))$, a family $\mathbf{F}$ of closed sets in X is defined by setting $A \in \mathbf{F}$ iff $\mathrm{Cl}(A) = A$.

(d) Given a family of closed sets $\mathbf{F}$ in X, a topological space $(X, \mathbf{G})$ is defined when we set $A \in \mathbf{G}$ iff the complementary set $(X - A) \in \mathbf{F}$.

(e) Given a topological space $(X, \mathbf{G})$, a neighborhood space (X, N^*) is defined when we set, for each x of X, $N^*(x) = [A$ in $P(X)$: A contains an H from $\mathbf{G}$ that contains the point $x]$.

(f) If we start with a neighborhood space (X, N) and follow through

all the earlier steps in order to come back to a neighborhood space (X, N^*), *then for any* x *of* X, $N^*(x) =$ [*the filter over* X *generated by the filter base* $N(x)$]. *And the two neighborhood spaces* (X, N) *and* (X, N^*) *lead to the same convergence space in step (b).*

PROOF. Given the definitions and the passages as outlined in Lemma 1.4.1, all the statements are easily verified. □

Lemma 1.4.2 *Given a set map* $f: X \to Y$ *and neighborhood spaces* (X, N), (Y, M), *the following statements are all equivalent, where the associated notions of convergence, closure, closed sets, or open sets are derived as in Lemma 1.4.1:*

(a) *for each point* x *of* X *and each* V *of* $M(f(x))$ *there is a* U *of* $N(x)$ *such that* $f(U) \subset V$;
(b) *whenever* $[(D, \leq), s, x]$ *is a convergence in* X, *then* $[(D, \leq), f \cdot s, f(x)]$ *is a convergence in* Y;
(c) *for any subset* A *of* X, $f[\mathrm{Cl}(A)] \subseteq \mathrm{Cl}[f(A)]$;
(d) *if* B *is any closed subset of* Y, *then* $f^{\mathrm{r}}(B)$ *is a closed subset of* X;
(e) *if* H *is an open subset of* Y, *then* $f^{\mathrm{r}}(H)$ *is an open subset of* X.

PROOF. The equivalence of these conditions is not hard to verify. When any, and so each, of these conditions is satisfied, we say that f is a *continuous map* of the first space X in the second space Y. For the neighborhood spaces we also define f to be *continuous at a* particular *point* x of X (only) if for each V from $M(f(x))$ there is a U of $N(x)$ such that $f(U) \subseteq V$.

The same set X may be the base for different neighborhood spaces $(X, N), (X, M), \ldots$. We talk then of *neighborhood structures* $N, M, \ldots$ for X. When N, M are neighborhood structures for X we say that N is *finer* than M (or M is *coarser* than N) if 1_X is continuous from (X, N) to (X, M). This is equivalent to saying that for each x of X the filter base $N(x)$ is finer than the filter base $M(x)$, or each set of $M(x)$ contains some set of $N(x)$. If each of M, N is finer than the other, we call N, M *equivalent* neighborhood structures for X; this is the same as saying that for each x of X, $N(x)$ and $M(x)$ generate the same filter over X. Correspondingly, we have various topologies ($\mathbf{G}, \mathbf{H}, \ldots$) for the same set X and talk of finer/equivalent topologies. Note that "$\mathbf{G}$ and $\mathbf{H}$ are equivalent" means $\mathbf{G} = \mathbf{H}$, whereas "$\mathbf{G}$ is finer than $\mathbf{H}$" means $\mathbf{G} \supseteq \mathbf{H}$. □

Lemma 1.4.3 *(a) The family of topologies on a set* X *forms a complete lattice* $(T(X), \leq)$ *under the order* $\leq$ *of "being finer than"; the zero element of this is* $[P(X)]$, *giving the discrete space* $(X, P(X))$; *the one element of the lattice is* $[X, \varnothing]$, *giving the indiscrete space* $(X, [X, \varnothing])$.

(b) Given a set mapping $f: X \to Y$ and a topology **H** *on Y, the family of those topologies* **G** *on X for which $f: (X, \mathbf{G}) \to (Y, \mathbf{H})$ is continuous forms a complete sublattice of T(X), with P(X) as least element and a greatest element denoted by $f^{\leftarrow}(\mathbf{H})$ (=[all $f^{r}(B)$: B in* **H**]).

(c) Given the set map $f: X \to Y$ and a topology **G** *on X, the family of those topologies* **H** *on Y for which $f: (X, \mathbf{G}) \to (Y, \mathbf{H})$ is continuous forms a complete sublattice of T(Y), with [Y, ∅] as its greatest element and a least element denoted by $f^{\rightarrow}(\mathbf{G})$ (=[all $B \subseteq Y$ for which $f^{r}(B)$ is in* **G**]).

PROOF. The easy proofs are left for the reader to complete. □

The space $(X, \mathbf{G})$ is called a *subspace* of the space $(Y, \mathbf{H})$ if X is a subset of Y, $\mathbf{G} = f^{\leftarrow}(\mathbf{H})$, for the inclusion map f of X in Y; so that $A \subseteq X$ is in **G** iff $A = X \cap B$ for some B of **H**.

The space $(Y, \mathbf{H})$ is the *quotient space* of the space $(X, \mathbf{G})$ *by the equivalence relation E* on X if $Y = X/E$ and $\mathbf{H} = p^{\rightarrow}(\mathbf{G})$ for the canonical surjective map p of X on $Y = X/E$. Thus a subset B of Y is in **H** iff the union of the E-classes that are the elements of B gives a set from **G**.

Finally, given an (indexed) family of topological spaces $[(X_j, \mathbf{G}_j)$: j in $J]$, the direct product of these spaces is defined as the space $(X, \mathbf{G})$, where $X = P[X_j$: j in $J]$, the product set, and **G** is the coarsest in the family of topologies $[\mathbf{G}_i]$ for X for which each of the canonical surjective maps p_j: $X \to X_j$, j in J, is a continuous map of $(X, \mathbf{G})$ in $(X_j, \mathbf{G}_j)$. This topology **G** is usually described by prescribing a neighborhood structure N^* for X under which $N^*(x) = P[A_j$: where A_j is a neighborhood of $p_j(x)$ in $(X_j, \mathbf{G}_j)$, and for all but a finite (or null) collection of the j's, $A_j = X_j]$. With this topology this product space is usually referred to as the *Tychonoff product* of the spaces given.

Given a set X, a family $(A_i$: i in $I)$ of subsets of X is called a *covering* of a subset A of X if A is contained in the union of the (A_i). Thus X itself is said to be covered by the subsets (A_i) if the union of the (A_i) is X itself. When we have a topological space $(X, \mathbf{G})$, a covering (A_i) of some A is called an *open covering* or a *closed covering* when all the sets A_i of the covering are open sets or closed sets, respectively, of the space. A family of subsets (B_i) of X is said to have the finite intersection property if the intersection of any finite selection of the B_i is nonnull.

In any space $(X, \mathbf{G})$ the following conditions are seen to be equivalent:

1. for each open covering (A_i) of X there is a finite subcovering $(A_{i(r)}$, $r = 1, \ldots, n)$; that is, a finite subfamily $(A_{i(r)}, r = 1, \ldots, n)$ of (A_i) is also a covering of X; and
2. if (B_i) is a nonnull family of closed sets of $(X, \mathbf{G})$ with the finite intersection property, then there is a point belonging to all the B_i.

(It is easy to prove the equivalence, since complementation takes open sets to closed sets and vice versa.)

A space $(X, \mathbf{G})$ is called a *compact space* if it satisfies either (and so each) of the foregoing conditions. A subspace that is compact as a space is called a *compact subset* of the space. The space $(X, \mathbf{G})$ is called a *locally compact space* if for each x of X the filter of neighborhoods at x has a base consisting of compact subsets of X. That is, each open set A containing x contains a compact set containing an open set containing x.

Given a topological space $(X, \mathbf{G})$, a filter base B over X is said to *converge to a point* x of X, and x is called a *limit point* of B in X, if the filter over X generated by B contains the filter of neighborhoods of x—in other words, if each open set containing x contains a set of B. The point x is called an *adherence point* of B if there is a filter containing B that converges to x—in other words, if each neighborhood of x meets each set of B (that is, has a nonnull intersection with it).

Lemma 1.4.4 *(a) A map $f: X \to Y$ is continuous from $(X, \mathbf{G})$ to $(Y, \mathbf{H})$ iff [a filter base B over X converges to a point x of X] implies [$f(B) = \{$the set of $f(A)$, A in $B\}$ is a filter base of Y that converges to $f(x)$ of Y].*

(b) A point x of X is an adherence point of a filter base B over X iff there is an ultrafilter containing B that converges to x.

(c) The following are equivalent: (i) the space $(X, \mathbf{G})$ is compact; (ii) each filter base B over X has an adherence point in X; (iii) each ultrafilter over X converges to some point of X.

PROOF. Part (a) follows from the definitions involved. For (b) it is enough to note that by use of Zorn's principle any filter base B over X has ultrafilters containing it. Now (b) implies that for an ultrafilter over X a limit point is the same as an adherence point.

(c) Assume that $(X, \mathbf{G})$ is compact: if B is a filter base over X, the family [Cl(A): A in B] is a family of closed sets with the finite intersection property from X. Hence there is a point x in all of them. That means that each neighborhood of x must meet each set of B, or that x is an adherence point of B. If next we assume that each filter base B over X has an adherence point in X, each ultrafilter, being a filter base, must have an adherence point; but then that is also a limit point of the ultrafilter. Finally, if we assume that each ultrafilter over X has a limit point in X, for any nonnull family of closed sets (B_i) of $(X, \mathbf{G})$ with the finite intersection property, they and their finite intersections constitute a filter base B. This base is contained in a ultrafilter that has a limit point x in X. Then x is in the closure of each set of B, and so in each of the sets B_i. Thus (X, **G**) is compact. □

Theorem 1.4.1 *(a) If f is a continuous map of $(X, \mathbf{G})$ in $(Y, \mathbf{H})$, then for any compact subset A of $(X, \mathbf{G})$, $f(A)$ is a compact subset of $(Y, \mathbf{H})$.*

(b) A closed subset A of a compact space $(X, \mathbf{G})$ is a compact subset.

(c) The product $P[(X_j, \mathbf{G}_j), j \text{ in } J]$ of any family of compact spaces $(X_j, \mathbf{G}_j)$ is also a compact space.

PROOF. (a) From the definition of a subspace it is clear that a subset A of $(X, \mathbf{G})$ is a compact subset iff each open covering of A (by open sets of $(X, \mathbf{G})$), has a finite subcovering. If we assume that A is a compact subset of $(X, \mathbf{G})$, then for any open covering (H_i) of the subset $f(A)$ of $(Y, \mathbf{H})$ the sets $[f^{\text{r}}(H_i)]$ form an open covering of A; hence a finite selection of these, say $[f^{\text{r}}(H_{i(r)}), r = 1, \ldots, n]$, already covers A. Then it is clear that the finite selection $[H_{i(r)}, r = 1, \ldots, n]$ of the (H_i) already covers $f(A)$. Hence $f(A)$ is a compact subset of $(Y, \mathbf{H})$.

(b) If A is a closed subset of the compact space $(X, \mathbf{G})$ and if (A_i) is any open covering of A, then $(A_i) \cup [(X - A)]$ is an open covering of X. Since $(X, \mathbf{G})$ is compact, a finite subcovering of X exists. But if this subcovering of X is $[A_{i(r)}, r = 1, \ldots, n] \cup [(X - A)]$, clearly the finite selection $[A_{i(r)}, r = 1, \ldots, n]$ covers A. Thus A is also a compact subset.

(c) If U is an ultrafilter over the product space $(X, \mathbf{G}) = P[(X_j, \mathbf{G}_j): j \text{ in } J]$ of a family of compact spaces $(X_j, \mathbf{G}_j)$, the images $p_j(U)$ of this family U under the canonical surjective maps of X on the X_j give filter bases over the X_j that generate ultrafilters over the X_j. Such an ultrafilter has a limit point x_j in $(X_j, \mathbf{G}_j)$, since that is a compact space. Then the point x of X with $p_j(x) = x_j$ for each j is a limit point of U: a typical neighborhood of x is a finite intersection of neighborhoods of the form $V_i^* = [P(A_j): A_j = X_j \text{ for all } j \neq i, A_i = \text{a neighborhood } V_i \text{ of } x_i = p_i(x)]$; since V_i contains a $p_i(A)$ for some A from U, V_i^* contains A from U. Hence a finite intersection of such V_i^* also would contain a set of U. Thus U converges to x. This proves that the product space is also a compact space. □

We next consider a series of *separation properties* for a space $(X, \mathbf{G})$:

0. $(X, \mathbf{G})$ is a *T_0-space* if $[x \neq y$ for points x, y of $X]$ implies that [some A of $\mathbf{G}$ exists that contains one of x, y but not both];
1. $(X, \mathbf{G})$ is a *T_1'-space* if for any x, y from X [some open set A from $\mathbf{G}$ contains x but not y] implies [some open set B from $\mathbf{G}$ contains y but not x];
2. $(X, \mathbf{G})$ is a *T_2'-space* if for any x, y from X [there exists an A from $\mathbf{G}$ that contains one of x, y but not both] implies [there exist disjoint sets U, V from $\mathbf{G}$ such that U contains x and V contains y];
3. $(X, \mathbf{G})$ is a *T_3-space* if [F is a closed set of $(X, \mathbf{G})$ and x is a point

not in F] implies [there exist disjoint sets U, V in **G** such that U contains F and V contains x];

4. (X, **G**) is a *T_4-space* if [F, K are disjoint closed sets of (X, **G**)] implies [there exist disjoint sets U, V in **G** such that U contains F and V contains K];
5. (X, **G**) is a *T_5-space* if [A, B are subsets of X such that $\mathrm{Cl}(A) \cap B = \varnothing$ and $\mathrm{Cl}(B) \cap A = \varnothing$] implies [there exist U, V in **G** that are disjoint such that U contains A and V contains B];
6. (X, **G**) is a *CR-space* (or completely regular space) if [F is a closed set of (X, **G**) and x is a point of X outside F] implies [there exists a continuous map f: (X, **G**) $\rightarrow$ the real unit interval I (with its usual topology) such that $f(x) = 0$ and $f(y) = 1$ for each y of F].

Lemma 1.4.5 *The following implications connect these separation properties: (a) $T_5 \Rightarrow T_4$; (b) (T_4 and T_1') $\Rightarrow$ CR; (c) CR $\Rightarrow$ (T_3 and T_2'); (d) (T_3 and T_1') $\Rightarrow T_2'$; and (e) $T_2' \Rightarrow T_1'$. Also, a compact T_2'-space is a T_4-space.*

For the proof of the lemma see Exercises and Remarks at the end of this chapter, page 26.

Finally, we collect (without proofs) results regarding connectedness in spaces.

1. A subset A of X is called a *connected subset* of the space (X, **G**) if [$A \subseteq B \cup C$ for B, C from **G**, and $B \cap A \cap C = \varnothing$] imply [either $B \cap A = \varnothing$ or $C \cap A = \varnothing$]. The space (X, **G**) itself is called a *connected space* if X is a connected subset of (X, **G**).
2. If [A_i: i in I] is a family of connected subsets of (X, **G**) such that they have at least one point contained in all of them, then their union is a connected subset of (X, **G**). Hence the union of all connected subsets of (X, **G**) containing a particular point x of X (which family surely contains the one-element set (x)) is the largest connected subset of (X, **G**) containing x. Calling it the *connected component* of (X, **G**) containing x, we note that two points are in some connected subset iff the connected components containing them are identical. These components are the maximal connected subsets of (X, **G**). There is only one such component iff the space is connected.
3. If f: (X, **G**) $\rightarrow$ (Y, **H**) is a continuous map, $f(A)$ is a connected subset of (Y, **H**) whenever A is a connected subset of (X, **G**). In particular, if f is a continuous map of the real unit interval I (as a connected space) in the space (X, **G**), the image set $f(I)$ is a connected set in (X, **G**); it is called *an arc* in (X, **G**) *connecting the point $f(0)$ with the point $f(1)$*. Two points p, q in (X, **G**) are said to be in the same "arcwise-connected component" of (X, **G**) if there is an arc from

p to q. Then it can be shown that the space partitions into maximal arcwise-connected components such that two points in different components cannot be connected by an arc, whereas those in the same component can be. When there is only one such component we call the space *an arcwise-connected space*. It would also be connected then.

4. A space $(X, \mathbf{G})$ is called *locally connected* if for each x in X the filter of neighborhoods at x has a base consisting of connected subsets of $(X, \mathbf{G})$.

Examples of the foregoing types are considered in Exercise 20 at the end of this chapter.

1.5 SEMIMETRIC AND SEMIUNIFORM SPACES

The familiar Euclidean n-space R^n with points that are (ordered) n-tuples of real numbers has a standard topology given by means of a metric or distance function d on it defined by $d(X, Y) = \sqrt{\Sigma(x_i - y_i)^2}$ if $X = (x_1, \ldots, x_n)$ and $Y = (y_1, \ldots, y_n)$ are any two points in the space. The topology is then prescribed by a neighborhood system N given by $N(X) = [S_e(X), e > 0]$ = the family of spheres about X as center and radii e taking all real positive values; the sphere $S_e(X)$ is defined as the set of points Y for which $d(X, Y) < e$. Such metric spaces have more structure than this associated topology or neighborhood structure, for we can talk of uniform continuity of functions between such spaces. The proper framework for this is the uniform or more general semiuniform space, and there is a corresponding more general version, the semimetric space.

Definition 1.5.1 A mapping $d: X \times X \to R$ (where R denotes the set of real numbers) is called (i) a *semimetric*, (ii) a *quasimetric*, or (iii) a *metric* on X when it satisfies the conditions (i) (D1) and (D2), (ii) (D1), (D2), and (D3), or (iii) (D1), (D2), (D3), and (D4), as stated next:

(D1) For all x, y from X, $d(x, x) = 0$ and $d(x, y) \geq 0$.
(D2) For all x, y, z from X, $d(x, y) + d(y, z) \geq d(x, z)$.
(D3) For all x, y from X, $d(x, y) = d(y, x)$.
(D4) $[d(x, y) = 0$ for x, y of $X]$ implies $[x = y]$.

Definition 1.5.2 Given a set X, the system $(U, (J, \leq), c)$ is called (i) a *semiuniformity*, (ii) a *quasiuniformity*, or (iii) a *uniformity* on X if: $(J, \leq)$ is a down-directed set, c is a map of J in itself, U is a monotone map of $(J, \leq)$ in the ordered set $(P(X \times X), \subseteq)$ subject to the conditions

(i) (U1), (U2), (ii) (U1), (U2), and (U3), or (iii) (U1), (U2), (U3), and (U4) below:

(U1) For each j of J, $U(j)$ contains 1_X.
(U2) Denoting by cj the image under c of the element j in J: for each j in J, $U(j)$ contains the relational product $U(cj) \circ U(cj)$.
(U3) For each j of J the relation $U(j)$ is symmetric; that is, $U(j)^r = U(j)$.
(U4) $[(x, y) \in \text{each } U(j)\text{: } j \text{ in } J]$ implies $[x = y]$.

Definition 1.5.3 We call (X, d) a *semimetric space*, a *quasimetric space*, or a *metric space* when d is a semimetric, a quasimetric, or a metric on X. And we call $(X, U, (J, \leq), c)$ a *semiuniform space*, a *quasiuniform space*, or a *uniform space* when $(U, (J, \leq), c)$ is a semiuniformity, a quasiuniformity, or a uniformity on X.

Definition 1.5.4 Given semiuniform spaces $(X, U, (J, \leq), c)$ and $(Y, V, (K, \leq), c)$, a mapping $f\colon X \to Y$ is called a *uniform map* of the first space in the second if for each k of K there is a j of J such that for all x, x' from X, $(x, x') \in U(j)$ implies $(f(x), f(x')) \in V(k)$. When $X = Y$ and $1_X\colon X \to X$ is a uniform map from $(X, U, (J, \leq), c)$ to $(X, V, (K \leq), c)$, we say that $(U, (J, \leq), c)$ is a *finer semiuniform structure* for X than $(V, (K, \leq), c)$. When each of the two semiuniform structures is finer than the other they are called *equivalent*. Also, when $f\colon X \to Y$ is a bijective map such that f and f^r are both uniform maps for the spaces $(X, U, (J, \leq), c)$ and $(Y, V, (K, \leq), c)$, f is called a *uniform equivalence* or a *unimorphism* between the spaces.

For given semiuniform spaces $(X, U, (J, \leq), c)$ and $(Y, V, (K, \leq), c)$ their *product* is defined as the semiuniform space $(X \times Y, U \times V, (J, \leq) \times (K, \leq), c)$ where $c(j, k) = (cj, ck)$ and $(U \times V)(j, k) = \{[(x, y), (x', y')]\colon (x, x') \in U(j), \text{ and } (y, y') \in V(k)\}$ for any (j, k) from $J \times K$.

Given a semimetric, quasimetric, or metric d on X, we have an associated semiuniformity, quasiuniformity, or uniformity on X, namely, $(S, (N^*, \geq), c^*)$ where N^* is the set of strictly positive natural numbers, $c^*(n) = n + 1$, and $S(n) = [(x, y)\colon d(x, y) < 1/2^n]$ for each n of N^*. We shall call this the standard semiuniformity defined by the semimetric d (or quasiuniformity defined by a quasimetric d or uniformity defined by a metric d). Again a semiuniformity $(U, (J, \leq), c)$ on X determines a neighborhood structure N, or $N(U)$, for X when we set $N(x) = [U_j(x)\colon j \text{ in } J]$, where y is in $U_j(x)$ iff (x, y) is in $U(j)$. The combination of these two steps leads, in the case of R^n with its metric d, to R^n with its usual topology. When is a uniformity defined as above by a metric? This is covered by the next theorem.

Theorem 1.5.1 *(a) A uniformity/quasiuniformity/semiuniformity* $(U, (J, \leq), c)$ *on a set* X *is derivable from a metric/quasimetric/semimetric* d *on* X *if and only if the filter over* $X \times X$ *generated by the filter base* $[U(j)$: j *in* $J]$ *has an enumerable base.*

(b) Any topological structure is derivable from a semiuniformity on X*; a topology* G *for* X *is derivable from a quasiuniformity (or a uniformity) if and only if the space* $(X, \mathbf{G})$ *is CR (CR and* T_0*).*

For the proof of the theorem see Exercises and Remarks at the end of this chapter, page 28.

1.6 COMPLETENESS AND THE CANONICAL COMPLETION

A semiuniformity $(U, (J, \leq), c)$ on X determines a *conjugate* semiuniformity $(U^r, (J, \leq), c)$ when we set $U^r(j)$ equal to the reverse of the relation $U(j)$ for each j of J; and these two conjugate semiuniformities also determine a quasiuniformity $(U^*, (J, \leq), c)$ with $U^*(j) = U(j) \cap U^r(j)$ for each j. U also determines a preorder relation $P = \cap[U(j)$: j in $J]$ and an equivalence relation $E = P \cap P^r = \cap[U^*(j); j \text{ in } J]$. Since E is contained in each $U(j)$ and in each $U^*(j)$, we also have associated semi- and quasiuniformities $(U/E, (J, \leq), c)$ and $(U^*/E, (J, \leq), c)$ on the quotient set X/E when we write, for A, B from X/E, $(A, B) \in (U/E)(j)$ [or $(A, B) \in (U^*/E)(j)$] iff for each x from A there is a y from B such that $(x, y) \in U(j)$ [$(x, y) \in U^*(j)$]. These quotient structures make X/E a T_0-space.

Given a down-directed set $(D, \leq)$ and a $(D, \leq)$-sequence s in X, we say that s is a *Cauchy* $(D, \leq)$*-sequence* in the semiuniform space $(X, U, (J, \leq), c)$ if for each j of J a nonnull initial set B of $(D, \leq)$ can be found such that $[s(b), s(b')]$ belongs to $U(j)$ for all b, b' from B. It is easy to see that when the $(D, \leq)$-sequence s in X converges to a point in the space $(X, T(U^*))$ then it is a Cauchy sequence in $(X, U, (J, \leq), c)$ where $T(U^*)$ is the topology on X determined by the quasiuniform structure $(U^*, (J, \leq), c)$. The converse need not be true in general. When it is we have a special type of space.

Definition 1.6.1 $(X, U, (J, \leq), c)$ is called a *complete semiuniform space* if for any down-directed set $(D, \leq)$ and any Cauchy $(D, \leq)$-sequence s in X there is a limit point for the sequence in $(X, T(U^*))$.

A filter base B over X is called a *Cauchy filter base* in $(X, U, (J, \leq), c)$ if for each j of J there is a set B in B such that $\mathrm{B} \times \mathrm{B} \subseteq U(j)$; we call such a set B *small of order* j. The role of these is given by the following:

Lemma 1.6.1 *(a) A Cauchy* $(D, \leq)$*-sequence* s *of* (X, U) *determines a Cauchy filter base* B *of* (X, U) *formed by the sets of the form* $[s(B)$: B

a nonnull initial subset of $(D, \leq)$*; and s converges to a point p in* $(X, T(U^*))$ *iff B converges to p in* $(X, T(U^*))$.

(b) A Cauchy filter base B over (X, U) *determines a* $(J, \leq)$*-Cauchy sequence s when we set s(j) to be some point chosen from some one set B(j) of B that is small of order J; and again B converges to P iff s converges to p.*

(c) Given a Cauchy filter base B over (X, U)*, a point p is an adherence point of B in* $(X, T(U^*))$ *iff it is a limit point of B.*

(d) A semiuniform space $(X, U, (J, \leq), c)$ *is complete iff either (i) each Cauchy filter base over* (X, U) *has a limit point in* $(X, T(U^*))$*, or (ii) each Cauchy* $(J, \leq)$*-sequence of* (X, U) *converges to a limit in* $(X, T(U^*))$.

PROOF. (a) Given the Cauchy $(D, \leq)$-sequence s of (X, U), for each j of J there is a d of D such that $(s(e), s(f))$ is in $U(j)$ for all $e, f \leq d$; hence there is a nonnull initial set $\Lambda(d)$ with $s(\Lambda(d))$ small of order j for each j of J. This shows that the family of sets [s(B): B a nonnull initial set of $(D, \leq)$] does form a Cauchy filter base B over (X, U). If s converges to a p of $(X, T(U^*))$, then for each j of J there is a d of D such that $s(\Lambda(d))$ is contained in $U^*(j)(p)$. Since this $\Lambda(d)$ is in B, B converges to p. Conversely, if B converges to a p of $(X, T(U^*))$, for each j of J there is first a B of B that is contained in $U^*(j)(p)$; and if d is in B, for each e of $\Lambda(d)$, e also belongs to B and $s(e)$ belongs to $U^*(j)(p)$. That is, s converges to p.

(b) Given a Cauchy filter base B over (X, U), we choose, for each j of J, a set B(j) in B small of order j, and then an element $s(j)$ from this B(j). This gives a Cauchy $(J,\leq)$-sequence s of (X, U), for when j is in J, for cj we can see that [$k, m \leq cj$ in J] implies that [$s(k)$, $s(m)$ being in sets B(k), B(m) of B that are small of order cj, there is q in B(k) and in B(m) so that $(s(k), q)$ and $(q, s(m))$ are both in $U(cj)$ and $(s(k), s(m))$ is in $U(j)$]. When B converges to a point p of $(X, T(U^*))$, given j of J and cj of J, there is a set B of B contained in $U^*(cj)(p)$. Then if $k \leq cj$, since B(k), B have a common element q, $(s(k), q)$ and (q, p) both are in $(U^*(cj)$, so that $(s(k), p)$ is in $U^*(j)$. This means that $U^*(j)(p)$ contains all $s(k)$ for $k \leq (cj)$, which in turn means that s converges to p in $(X, T(U^*))$. Conversely, if s converges to p, given j in J and c^2j in J, $U^*(c^2j)(p)$ contains all $s(k)$ for k less than or equal to a certain k_0 of J; if B is a set of B small of order (c^2j) and m an element of J that is less than or equal to both (c^2j) and k_0 (such elements exist since $(J, \leq)$ is down-directed), then $s(m)$ belongs to B(m) $\cap$ $U^*(c^2j)(p)$. Hence if q is a point common to the sets B(m), B (from B), then $(p, s(m))$, $(s(m), q)$ and (q, r) all belong to $U^*(c^2j)$ for any r from B; thus (p, r) is in $U^*(j)$ for all r of B, which means that $U^*(j)(p)$ contains this B from B. Hence B converges to p.

(c) Part (i) follows from (a) and (b) since [each Cauchy filter base over (X, U) converges] iff [each Cauchy sequence over (X, U) converges].

For part (ii) the "only if" part is trivially true. If we assume then that each Cauchy $(J, \leq)$-sequence of (X, U) has a limit in $(X, T(U^*))$, then by part (b) each Cauchy filter base over (X, U) has a limit in $(X, T(U^*))$. By the part (i) just proved (X, U) is complete. □

We now describe, for any semiuniform space, an associated complete semiuniform and T_0-space called its *canonical completion*:

Theorem 1.6.1 *Given a semiuniform space $(X, U, (J, \leq), c)$, there is an associated complete T_0-semiuniform space $(X\#, U\#, (J, \leq), c\#)$ called its canonical completion with the following properties: (i) there is a uniform mapping h of $(X, U, (J, \leq), c)$ in $(X\#, U\#, (J, \leq), c\#)$ such that $(x, y) \in U(cj)$ implies that $(h(x), h(y)) \in U\#(j)$ for each j of J; and (ii) if there is a uniform map g of $(X, U, (J, \leq), c)$ in any complete T_0-semiuniform space $(Y, V, (K, \leq), c)$, there is a unique uniform map $g\#$ of $(X\#, U\#, (J, \leq), c\#)$ in $(Y, V, (K, \leq), c)$ such that $g = g\# \cdot h$. If the original space $(X, U, (J, \leq), c)$ has a (i) quasiuniformity, or (ii) a semiuniformity defined by means of a semimetric, quasimetric, or metric on X, then $(U\#, (J, \leq), c\#)$ is a similar type of semiuniformity on $X\#$.*

For the proof of the theorem see Exercises and Remarks, page 30.

EXERCISES AND REMARKS ON CHAPTER 1

1. Given a relation θ on A, verify that there is (a) a smallest reflexive relation containing θ (namely, $\theta \cup 1_A$), (b) a smallest symmetric relation containing θ (namely, $\theta \cup \theta'$), (c) a smallest transitive relation containing θ (namely, $\cup[\theta^n: n = 1, 2, \ldots]$, where $\theta^1 = \theta$ and $\theta^{n+1} = \theta \circ \theta^n$ for $n = 1, 2, \ldots$), and (d) a smallest equivalence relation containing θ (namely, $\cup[\omega^n: n = 1, 2, \ldots]$, where $\omega = \theta \cup 1_A \cup \theta'$).

2. When the product relation $\sigma \circ \rho$ is defined, show that it is injective, surjective, bijective, a mapping relation, or bi-injective when each of ρ and σ is such a relation.

3. For the power set $P(X)$ consisting of all subsets of a set X, prove the following properties of the complement operation C: for subsets A, B, and A_i, all of X, $C(C(A)) = A$; $A \cup C(A) = X$; $A \cap C(A) = \varnothing$; $A \subseteq B$ implies $C(B) \subseteq C(A)$, $C[U(A_i)] = \cap[C(A_i)]$.

4. Check that (X, θ) is a semigroup in which each element is both a left unit and a right zero if we define $x \, \theta \, y = y$ for each pair of elements x, y from

X. Show that this semigroup can have no right unit (unless X has just one element).

5. Show that a semigroup (X, θ) with a left unit element e gives rise to a group $(X, \theta, (\ )')$ if the unary operation on X is so defined that $(x)' \theta x = e$ for each x of X; that is, if each x of X has a left-inverse $(x)'$ relative to e. (*Hint*: Prove that there is left-cancellation for each element of X; then show that the mapping h defined in Exercise 8(c), below, is into $BM(X)$ and it takes $(\ )'$ to the reverse relation.)

6. In a semigroup, y is called a relative inverse of x if $x \theta y \theta x = x$ and $y \theta x \theta y = y$; thus the semigroup is a regular semigroup if each element has a relative inverse. Show that for x, if there is a y such that $x \theta y \theta x = x$, then x has a relative inverse (namely, $y \theta x \theta y$). Also show that in an inverse semigroup each element has a unique relative inverse. (See Ljapin [21], Theorem 7.4.)

7. If (X, θ) is a semilattice, an order relation $\leq$ can be defined on X such that for each pair of elements x, y of X there is a greatest lower bound (or g.l.b) under the order, namely, $x \theta y$ and conversely. (See Birkhoff [2], Chapter 1, Lemma 1 and Corollary.)

8. PROOF OF THEOREM 1.2.2. (a) An element α of $PM(A)$ can be viewed as a mapping of dom(α) in ran(α); with this α we consider also the equivalence relation $E(\alpha)$ defined on dom(α) by $E(\alpha) = \alpha^r \circ \alpha$; to each subset C_i of dom(α) containing exactly one element from each of these equivalence classes of dom(α) under $E(\alpha)$, we can associate a family of elements (α_i'), all from $M(A)$: a typical such α_i' is of the form $[(b, a)$: with one (b, a) for each b of A, with a from C_i, and with (a, b) in α when b is in ran(α)]. All these are relative inverses of α in $PM(A)$, and in $M(A)$ when α is in $M(A)$. Other relative inverses of α from $PM(A)$ can be obtained from such an α_i' by taking elements of the form $\alpha_i' \circ 1_D$, where D is any subset of A containing ran(α). Surely 1_A is a unit element in all these semigroups.

(b) When we start with a bi-injective relation α on A, the last-defined $E(\alpha)$ becomes the identity relation, the set C_i is unique, and the inverses α_i' are only those that contain α^r; this α^r is the unique relative inverse (also belonging to $BI(A)$) that we now use. To see that $(BI(A), \circ, (\ )')$ is an inverse semigroup, note that the typical idempotent element $\alpha \circ \alpha^r$ or $\alpha^r \circ \alpha$ is now $1_{\text{ran}(\alpha)}$ or $1_{\text{dom}(\alpha)}$; and for any C, D (subsets of A), $1_C \circ 1_D = 1_{C \cap D} = 1_D \circ 1_C$; thus all idempotents in this semigroup commute. And for a bijective mapping α on A, α^r is a bijective mapping too, and the relative inverse of α. In $BM(A)$ an idempotent is of the form $\alpha \circ \alpha^r = 1_A$, so that this has only one idempotent. It is also a unit element. Thus $(BM(A), \circ, (\ )')$ is a group.

(c) Starting from a given semigroup (X, θ), we can construct a groupoid homomorphism h: $(X, \theta) \to (M(X), \circ)$ by setting $h(x) = [(y, x \theta y)$: y in $X]$. Since θ is a binary operation on X, each $h(x)$ is a mapping relation on X, that is, belongs to $M(X)$; and from the associativity of θ we get $h(y) \circ h(x) = h(y \theta x)$, so that h is a homomorphism. If the semigroup (X, θ) has a unit element e or even just a left-cancelable element e (for which $e \theta x = e \theta y$

implies $x = y$), h becomes an injective map. For if $h(x) = h(y)$, this mapping relation must contain both $(e, e \theta x)$ and $(e, e \theta y)$, so that $e \theta x = e \theta y$, and finally $x = y$. Thus h is an isomorphism of (X, θ) with a subsemigroup of $(M(X), \circ)$ in this case.

If we start as above, but with a regular semigroup $(X, \theta, (\ )^*)$, the same h gives a groupoid homomorphism and an isomorphism with a subgroupoid in $(M(X), \circ)$ for (X, θ). For an x of X, if $h(x) = \alpha$ in $M(X)$, under the equivalence $E(\alpha)$ we see that $(y, x^* \theta x \theta y)$ always belongs to $E(\alpha)$ for each y of X. We can then take a C_i to consist of these special elements $(x^* \theta x \theta y$: y in $X)$, one from each $E(\alpha)$ class. A relative inverse for α is then $h(x^*)$ itself. Thus h becomes a regular groupoid homomorphism and isomorphism in $(M(X), \circ, (\ )')$.

The case of the group is treated similarly; the same h as defined earlier now gives an $h(x)$ that is a bijective mapping of X on X, for $x \theta y = x \theta z$ implies $x^* \theta x \theta y = x^* \theta x \theta z$ or $1 \theta y = 1 \theta z$, or $y = z$; and for any z of X there is a y of X such that $x \theta y = z$, namely, $y = x^* \theta z$. Further, it is clear that now $h(x^*)$ is the reverse of the relation $h(x)$: $h(x^*) = h(x)^r$. Thus h is an isomorphism of $(X, \theta, (\ )^*)$ with a subregular groupoid (or subgroup) of $(BM(X), \circ, (\ )')$.

We treat the case of the inverse semigroup somewhat differently. We assume the given inverse semigroup is $(X, \cdot, (\ )^*)$ with the binary operation written as a multiplication dot. We further simplify the notation and write just xy for $x \cdot y$ and x^* for $(x)^*$. We define the mapping g: $(X, \cdot, (\ )^*) \to (BI(X), \circ, (\ )')$ by setting, for any x of X, $g(x) = [(x^*t, xx^*t)$: t in $X]$. The domain of $g(x)$ is x^*X, its range is xX, for any xs of xX can be written as $xx^*(xs)$. Further, $g(x^*) = [(xt, x^*xt)$: t in $X] = [(x^*s, xx^*s)$: s in $X]^r$ by writing $s = xt\ (= xx^*s)$. That is, $g(x^*) = [g(x)]^r$. We can also check that $g(x) \circ g(y) = g(xy)$: for $(p, q) \in g(x) \circ g(y) \Rightarrow$ [for some s of X, $p = y^*s$; for some t of X, $x^*t = yy^*s$ and $q = xx^*t] \Rightarrow [p = y^*s = y^*yy^*s = y^*x^*t$ and $q = xx^*t = xyy^*s = xyy^*yy^*s = xyy^*x^*t] \Rightarrow [(p, q) \in g(xy)]$; and conversely, $(p, q) \in g(xy) \Rightarrow$ [for some t of X, $p = y^*x^*t$ and $q = xyy^*x^*t] \Rightarrow$ [for $s = yy^*x^*t$, $p = y^*s$, $yy^*s = yy^*yy^*x^*t = yy^*x^*t = s = yy^*x^*xx^*t = x^*xyy^*x^*t = x^*xs$, and $xx^*xs = xs = q] \Rightarrow [(p, q) \in g(x) \circ g(y)]$. Thus g is indeed a homomorphism of the regular groupoid and inverse semigroup $(X, \cdot, (\ )^*)$ in $(BI(X), \circ, (\ )')$. We need only show that g is an injective mapping to prove that g is an isomorphism of the given inverse semigroup with a subregular groupoid of $(BI(X), \circ, (\ )')$. Suppose that $g(x) = g(y)$. Then $g(x^*) = [g(x)]^r = [g(y)]^r = g(y^*)$; from $(x^*, xx^*) \in g(x) = g(y)$ we deduce that for some t of x, $x^* = y^*t$ and $xx^* = yy^*t = yx^*$. By a similar argument we can show that $yy^* = xy^*$; and using $g(x^*) = g(y^*)$, by similar arguments we get $x^*x = y^*x$ and $y^*y = x^*y$. Hence we now have $x = (xx^*)x = (y)x^*x = y(y^*y)(x^*x) = yx^*(xy^*)y = yx^*(yy^*y) = y(x^*y) = yy^*y = y$. Hence g is injective, as we wished to show.

9. Characterize a group as an FD-algebra, with F containing a binary, a unary, and a nullary operation each, and D consisting of suitable pairs from $P(A, F)$, where A is the set $[a, b, c]$.

10. Establish a one-to-one correspondence $N \leftrightarrow C$ between the family of normal subgroups of a group and the family of congruences on it.

11. When f: $(X, F) \to (Y, F)$ is a surjective F-homomorphism, show that there is a one-to-one correspondence $E^* \leftrightarrow E$ between the family of congruences (E^*) on (X, F) that contain the relation $f^{\mathrm{r}} \cdot f$ and the family (E) of congruences on (Y, F) such that when E^*, E are associated, $(X/E^*, F)$ and $(Y/E, F)$ are isomorphic.

12. Among rings with unit, the ring $(Z, +, -, 0, \cdot, 1)$ has a generating set [1] consisting of just one element. Deduce that for any ring $(R, +, -, 0, \cdot, 1)$ with unit element, and for any given element x of R, there is a unique homomorphism of the ring $(Z, +, -, 0, \cdot, 1)$ in the given ring which maps 1 on x.

13. Given a nonnull set A, we call any finite sequence $(a_1, \ldots, a_n)$ of elements from A a *word over* A. Defining a composition $\circ$ for such words $(a_1, \ldots, a_n)$ and $(b_1, \ldots, b_m)$ by $(a_1, \ldots, a_n) \circ (b_1, \ldots, b_m) = (a_1, \ldots, a_n, b_1, \ldots, b_m)$, show that these words (including a null word with no letters) form a semigroup under this composition, and that this semigroup $(W(A), \circ)$ is isomorphic with the "free semigroup with 1" over the set A.

14. Work out proofs of Lemmas 1.4.1, 1.4.2, and 1.4.3 and parts (a) and (b) of Lemma 1.4.4.

15. Prove that the following conditions are all equivalent for a filter F over X and that they characterize an ultrafilter over X: (i) [$A \cap B$ is nonnull for each B from F] implies [A belongs to F]; (ii) for each subset A of X, either A or $X - A$ belongs to F; (iii) [$A \cup B$ belongs to F] implies [A or $B \in F$].

16. If f: $(X, \mathbf{G}) \to (Y, \mathbf{H})$ is continuous and $\mathbf{G} = f^{\leftarrow}(\mathbf{H})$, show that (a) if $(X, \mathbf{G})$ satisfies any one of the conditions T_0, T_1', T_2', T_3, CR, or T_4, so does $(Y, \mathbf{H})$; and (b) if $(Y, \mathbf{H})$ satisfies any of the conditions T_1', T_2', T_3, CR, or T_4, so does the space $(X, \mathbf{G})$. If in the space $(X, \mathbf{G})$ the equivalence E is defined as follows: xEy iff every open set containing one of x, y contains both, then show that the space $(X/E, \mathbf{H})$, where $\mathbf{H}$ is the topology $p^{\rightarrow}(\mathbf{G})$ associated with the canonical surjection p of X on X/E, is a T_0-space.

17. Show that a subspace of a direct product of spaces each of which is T_0, T_1', or T_2' also satisfies the same condition.

18. *Example 1.* The set R of all real numbers is given a neighborhood structure N by setting $N(x) = \{(y: y < x + \epsilon): \text{for } \epsilon \text{ real} > 0\}$; show that this space is T_0, but not T_1' nor T_2'.

Example 2. The usual real unit interval I is augmented with an extra point $0'$ that has the same neighborhoods as 0; that is, for both 0 and $0'$ a typical neighborhood is of the form $(0') \cup [0, \epsilon)$ for $\epsilon > 0$. This space can be shown to be not T_0, but it is compact T_4, and therefore CR, T_3, T_2', and T_1'.

Example 3. A space is now obtained from the unit interval by changing the neighborhood system at the points 0 and 1 only; the base of neighborhoods at 0 is of the form $\{[0, e) \cup (1 - e, 1)$: for $0 < e < \frac{1}{2}\}$, and at 1 it is of the form $\{(0, e) \cup (1 - e, 1]$: for $0 < e < \frac{1}{2}\}$. This space is T_1', not T_2'.

Example 4. This time the neighborhoods at 0 alone are changed to be the sets of the form $\{[0, e) \cap A\}$ where A is the complement in I of the set of numbers $1, \frac{1}{2}, \frac{1}{3}, \ldots, 1/n, \ldots$. This space is T_0, T_2', but not T_3.

Example 5. The subspace$[(x, y): y \geq 0]$ of R^2 is modified by changing the neighborhood system at the points $(x, 0)$ on the x-axis alone; for such a point a typical neighborhood is taken to be of the form $[(x, 0) \cup \{(x', y')$: for $(x' - x)^2 + (y' - e)^2 < e^2\}]$, which is an open circle tangent to the x-axis and above it, touching it at the point $(x, 0)$, together with the point $(x, 0)$. There are neighborhoods like this for each real $e > 0$. This space can be seen to be T_3 but not T_4: the two sets $[(x, 0)$ with x rational$]$ and $[(x, 0)$ with x irrational$]$ are disjoint closed sets of the space for which disjoint open sets containing them cannot be found.

19. Show that in a (T_0 and T_2')-space, or what is usually called a *Hausdorff space*, any filter can have at most one limit point.

20. Show that an arcwise-connected space is a connected space.

Example 6. This is a space that is connected, but not arcwise connected nor locally connected: it is the subspace of R^2 defined on the subset $A = [$(the y-axis) $\cup$ (the points of $y = \sin(1/x)$ for $x > 0$)$]$.

Example 7. This gives an arcwise-connected space that is not locally connected: it is the subspace of the last example together with the points on the x-axis.

21. PROOF OF LEMMA 1.4.5. We leave to the reader the easy proofs of parts (a), (c), (d), and (e).

(b) Let us then start with a (T_4 and T_1')-space $(X, \mathbf{G})$ and a closed set F in it and a point x outside F. Since there is an open set $(X - F)$ containing x but not containing y, for any y of F, by condition T_1' there is an open set $O(y)$ containing y but not containing x; this being true for each y of F, none of them can belong to $\mathrm{Cl}(x)$. Thus $\mathrm{Cl}(x)$ and F are disjoint closed sets of $(X, \mathbf{G})$. We rename these $F(0)$ and $F(1)$; since the space is T_4, there exist disjoint open sets containing $F(0)$, $F(1)$, respectively. We name these $G(\frac{1}{2})$ and $H(\frac{1}{2})$, and set $G(0) = X - F(1)$ and $H(0) = X - F(0)$; if we define the open sets $O(0, \frac{1}{2}) = G(\frac{1}{2}) - F(0)$, and $O(\frac{1}{2}, 1) = H(\frac{1}{2}) - F(1)$, corresponding to the three binary numbers $(\frac{0}{2}, \frac{1}{2}, \frac{2}{2})$ with denominator $2 = 2^1$, we have a partition of the space X as the union of $[F(0), O(0, \frac{1}{2}), F(\frac{1}{2}), O(\frac{1}{2}, 1), F(1)]$ with the O's as open sets and the F's as closed sets. We get a second such partition corresponding to the binary numbers with denominator 2^2 as follows. The disjoint closed sets $F(0)$, $X - G(\frac{1}{2})$ possess disjoint open sets containing them, respectively; we choose such sets and call them $G(\frac{1}{4})$ and $H(\frac{1}{4})$. Similarly, for the disjoint closed sets $X - H(\frac{1}{2})$ and $F(1)$ we choose disjoint open sets containing them and denote these by $G(\frac{3}{4})$ and $H(\frac{3}{4})$, respectively. Then if we define $O(0, \frac{1}{4}) = G(\frac{1}{4}) - F(0)$, $O(\frac{1}{4}, \frac{1}{2}) = H(\frac{1}{4}) \cap G(\frac{1}{2})$, $O(\frac{1}{2}, \frac{3}{4}) = H(\frac{1}{2}) \cap G(\frac{3}{4})$, and $O(\frac{3}{4}, 1) = H(\frac{3}{4}) - F(1)$, we get a new partition of X: $[F(0), O(0, \frac{1}{4}), F(\frac{1}{4}), O(\frac{1}{4}, \frac{1}{2}), F(\frac{1}{2}), O(\frac{1}{2}, \frac{3}{4}), F(\frac{3}{4}), O(\frac{3}{4}, 1), F(1)]$, which is a refinement of the first, and contains the F-sets of the earlier partition. This can be continued for binary numbers with denominators 2^3, then for those with denominators 2^4, and so on. Now we can define the function f: $X \to I$ in terms of these partitions by setting $f(y) = r$ if y is in the closed set $F(r)$, with r of the form $(a/2^m)$; otherwise, y would be in a decreasing sequence of open sets of the form $O(r_1, s_1)$, $O(r_2, s_2)$, ... where the pairs of rationals (r_1, s_1), (r_2, s_2), ... form a nest of intervals in I of diameters $\frac{1}{2}$, $\frac{1}{4}$, ... that then determines a unique real number that lies in each of the intervals. This real number is taken to be $f(y)$ in this case. It can be shown that this gives indeed a continuous map of $(X, \mathbf{G})$ in I, and surely $f(x) = 0$ and $f(y) = 1$ for y in F. This proves part (b).

To prove the last statement, assume that $(X, \mathbf{G})$ is a T_2' and compact space. If F, K are disjoint closed subsets of the space, they are also compact subsets in X. For any p in F and any q in K, since $X - F$ is an open set that contains q but not p, there must be disjoint open neighborhoods $U_q(p)$, $V_p(q)$, since the space is T_2'. Choosing such a pair for a fixed p of F and each of the q's in K, we get an open covering $[V_p(q)]$ of the compact set K. Hence from this we can select a finite subcovering. That is, we can find $q_1, \ldots, q_n$ (from K) such that $K \subseteq \cup[V_p(q_j): j = 1, \ldots, n]$, while p is contained in the open set $\cap[U_{q_j}(p): j = 1, \ldots, n]$. These sets named as $G_p(K)$ and $H(p)$ are disjoint open sets containing K and p. Having chosen such a pair of disjoint open sets, for each p of F, we get an open covering $[H(p)$: p in $F]$ of the compact set F. From this then we can choose a finite subcovering, say $[H(p_i): i = 1, \ldots, m]$. Then $H = \cup[H(p_i)]$ and $G = \cap[G_{p_i}(K)]$ give a pair of disjoint open sets containing F and K, respectively. Thus T_4 has been proved. □

22. On compact T_2' semiuniform spaces, the following conditions hold:

(a) Any compact T_2'-space $(X, \mathbf{G})$ has a semiuniformity $(U, (\mathbf{C}, \leq), i)$ defining its topology, where $\mathbf{C}$ is the family of finite open coverings of X, $\leq$ denotes refinement ($C \leq C'$ means that C is a refinement of C'), and $i(C) = C$. Since any two finite coverings have a common refinement, which is also a finite covering, $(\mathbf{C}, \leq)$ is indeed a down-directed set. We define $U(C) = [(x, y) \in X \times X$: ($x \in$ a set A of C) implies that ($y \in A$)]. This is surely then a semiuniformity for X. To see that the topology it defines is $\mathbf{G}$, note that $U(C)(x)$ is the open set that is the intersection of all the A of C that contain x, and that when x is in an open set G, we can cover $X - G$ by a finite family of open sets that, with G, provide a C such that $U(C)(x)$ is contained in G.

(b) When f: $(X, T(U)) \to (Y, T(V))$ is a continuous map for a compact semiuniform space (X, U) and a semiuniform space (Y, V), then f is also a uniform map. Once this is proved, it can be shown that there is only one semiuniformity [that given in (a)] that gives the original topology for a compact T_2'-space (X, G), for the possible semiuniformities form a complete lattice (under the relation of "finer than"); then the finest of them must be equivalent to each of the others

(c) A semiuniform space $(X, U, (J, \leq), c)$ is called *totally bounded* if for each j of J there is a finite covering of X by open sets small of order j. With this definition it can be shown that a semiuniform space (X, U) is complete and totally bounded iff the space $(X, T(U))$ is a compact space.

23. PROOF OF THEOREM 1.5.1. (a) The "only if" part follows from the definitions. Assume then that the filter generated by the $[U(j)]$ has an enumerable base $[B_n: n = 1, 2, \ldots]$. We can replace this by another enumerable base $[V_n]$ such that for each n, V_n contains $V_{n+1} \circ V_{n+1} \circ V_{n+1}$. In terms of these V we first define a mapping $g: X \times X \rightarrow I$ by $g(x, y) = 0$ if (x, y) is in each V_n; $g(x, y) = 1$ if (x, y) is outside even V_1, and $g(x, y) = 1/2^n$ if (x, y) is in V_n but not in V_{n+1}. Clearly this g is defined uniquely and well for each (x, y) from $X \times X$, and $(x, y) \in V_n$ iff $g(x, y) \leq 1/2^n$, or $g(x, y) < 1/2^{n-1}$. Now we can define $d: X \times X \rightarrow I$ by $d(x, y) = \inf[\Sigma g(z_j, z_{j+1})$; these sums taken over all finite sequences of points $x = z_0, \ldots, z_n = y$ from x to $y]$. Surely then $d(x, y) = 0$, $d(x, y)$ lies in I for each pair (x, y), and for any triple points x, y, z, $d(x, y) + d(y, z) \geq d(x, z)$, so that d is a semimetric on X; and it is seen that d is a quasimetric or metric when the semiuniformity $(U, (J, \leq), c)$ is a quasiuniformity or uniformity. So we have only to check that this semimetric determines essentially the same semiuniformity as that we started with. For this we first prove the following:

$$\forall x, y \text{ from } X,\ 1/2g(x, y) \leq d(x, y) \leq g(x, y); \tag{1.5.1}$$

the second statement is evident from the definitions. To prove that $d(x, y) \geq 1/2g(x, y)$, it is sufficient to show that for any finite sequence $x = z_0, \ldots, z_n = y$ from x to y, the sum $\Sigma\ g(z_i, z_{i+1})$ is $\geq 1/2g(x, y)$. This we prove by induction on the length n. Clearly it is true for $n = 1$; if it has been proved for all m up to $(n - 1)$, where $n \geq 2$, we next prove it for n; the result is trivially true if $g(x, y) = 0$, so we assume $g(x, y) > 0$. Taking the sequence $z_0, \ldots, z_n$ from x to y of length n, we let s denote the sum $\Sigma g(z_i, z_{i+1})$; again if $s = 0$, this would imply that $g(x, y) = 0$, so s must be >0. Let us then choose the smallest integer h such that $g(z_0, z_1) + \cdots + g(z_h, z_{h+1})$ exceeds $s/2$ (which is >0); h would be an integer between 0 and $(n - 1)$. We split the argument into three cases.

Case 1: $h = 0$. This means that $g(z_0, z_1)$ is $\geq s/2$; hence $g(z_1, z_2) + \cdots + g(z_{n-1}, z_n) < s/2$; if t is the largest integer such that $1/2^t \leq s$, we have $g(z_0, z_1) < s$, and by the induction hypothesis, $1/2 \cdot g(z_1, z_n) < s/2$, giving $g(z_1, y) < s$ also. Since these are numbers of the form $1/2^r$, each of them must be $\leq 1/2^t$; hence (x, z_1) and (z_1, y) both belong to V_t. Then (x, y) belongs to $V_t \circ V_t \leq V_{(t-1)}$. That is, $g(x, y) < 1/2^{t-1} = 2/2^t \leq 2s$, since $1/2^t \leq s$. This finally gives $1/2 \cdot g(x, y) \leq s$.

Case 2: $h = n - 1$. This means that $g(x, z_1) + \cdots + g(z_{n-2}, z_{n-1}) < s/2$ and $g(z_{n-1}, y) < s$; as before, the induction hypothesis gives $g(x, z_{n-1}) < s$ also. Hence again we get $1/2 \cdot g(x, y) \leq s$.

Case 3: $0 < h < n - 1$. There are three parts in the sum, such that the first and third have a value $<s/2$ while the middle one is $<s$; as before, the induction hypothesis then gives three terms: $g(x, z_h)$, $g(z_h, z_{h+1})$, and $g(z_{h+1}, y)$, each of them $<s$, and so $\leq 1/2^t$. Then $(x, y) \in V_{(t-1)}$, since each of the three (x, z_h), (z_h, z_{h+1}), (z_{h+1}, y) is in V_t. This again gives finally $1/2 \cdot g(x, y) \leq s$. This completes the induction proof.

If now $(S, (N^*, \geq), c^*)$ is the semiuniformity defined on X by the semimetric d, the relations (1.5.1) between g and d imply that $V_{n+1} \subseteq S(n) \subseteq V_n$; since the (V_n) form a base for the filter generated by the $[U(j)]$, so do the $[S(n)]$. That is, the semiuniformity defined by d is equivalent to the original one with which we started.

(b) Given a topology **G** on X, to each A from **G** associate a subset A^* of $X \times X$ by $A^* = [(x, y): x \text{ in } A \text{ implies } y \text{ in } A] = (A \times A)U[(X - A) \times X]$. For the $U(j)$ we take the finite intersections of these A^*; taking $c(j) = j$, we do then get a semiuniformity for X that determines the topology **G**. □

Before proving the other parts of this (b) we pause to note a corollary to what was proved under (a), namely:

Corollary *Any semi- (or quasi)uniformity on X can be expressed as a lattice product of semi- (quasi)metric-defined semi- (quasi)uniformities on the same basic set of poirts. For if the semi- or quasiuniformity is $(U, (J, \leq), c)$, this family $[U(j)]$ can be obtained as the union of a family $[U_k^*: k \text{ in } K]$ of enumerable subfamilies, with each U_k^* of the form $(U(j), U(c^2j), U(c^4j), \ldots)$ for some j of J, and as in the proof of part (a), each of these U_k^* is equivalent to the semi(quasi)uniformity on X defined by an associated semi(quasi)metric d_k on X. Hence the original semi(quasi)uniform structure is the lattice product of these defined by semi(quasi)metrics on X.*

If we now have the space $(X, \mathbf{G})$ defined by the quasiuniformity $(U, (J, \leq), c)$ on X, and p is a point in X, F a closed set not containing p, then $(X - F)$ being an open set containing p, $(X - F)$ must contain some $U_j(p)$, j in J. Let d be the quasimetric on X associated with the sequence $(U(j), U(c^2j), U(c^4j), \ldots)$, and set $f(y) = d(x, y)$ for any y of X. It can be seen that this f is a continuous map of $(X, \mathbf{G})$ in I, with $f(p) = 0$ and $f(y) = 1$ for y in F. Thus $(X, \mathbf{G})$ is a completely regular space. If we had started out with a uniformity $(U, (J, \leq), c)$, then the associated space $(X, \mathbf{G})$ would also be a T_0-space.

Conversely, given a completely regular space $(X, \mathbf{G})$, consider the family of pairs (p, K) of a point p of X and a closed set K not containing p. For each such pair $k = (p, F)$ we choose an associated continuous map f_k of $(X, \mathbf{G})$ in I such that $f_k(p) = 0$ and $f_k(y) = 1$ for y in K. This in turn determines a quasimetric d_k on X by $d_k(x, y) = |f_k(y) - f_k(x)|$; it can be shown then that the lattice product of the quasiuniformities defined by these several quasimetrics is a quasiuniformity on X that determines the topology **G**. If the space $(X, \mathbf{G})$ were also T_0 besides being completely regular, the foregoing quasiuniformity would be a uniformity. (See Krishnan [19].)

24. PROOF OF THEOREM 1.6.1. Starting with the given semiuniform space $(X, U, (J, \leq), c)$, we first construct a complete semiuniform space $(X', U', (J, \leq), c)$ over the set X' = [the family of all Cauchy $(J, \leq)$-sequences from (X, U)]. With $(J, \leq)$ and c as before, we define $U'(j) = [(x', y') \in X' \times X'$: there exists a nonnull initial set B of $(J, \leq)$ such that $(x'(k), y'(k)) \in U(j)$ for each k in B]. Since $j \leq k$ implies that $U(j) \subseteq U(k)$, it also implies $U'(j) \subseteq U'(k)$; surely (x', x') is in each $U'(j)$ for any x' of X', since x' is a Cauchy $(J, \leq)$-sequence from (X, U). And finally, for x', y', z' from X', if (x', y') and (y', z') are both in $U'(cj)$, then (x', z') is in $U'(j)$ follows, since the intersection of two nonnull initial subsets of $(J, \leq)$ is also such a set. Hence $(U', (J, \leq), c)$ does provide a semiuniform structure for X'. To see that this gives a complete space: we shall show that any Cauchy $(J, \leq)$-sequence s' in (X', U') has a limit point z' in $(X', T(U'^*))$. To get to this z' we first choose, for each j of J, a j' in J such that $[k, m \leq j']$ implies that $[s'(k), s'(m)) \in U'^*(c^2j)]$, and then choose a j'' such that in the Cauchy $(J, \leq)$-sequence $s'(j')$ of (X, U), $[r, s \leq j'']$ would imply that $[(s'(j')(r), s'(j')(s)) \in U^*(c^2j)]$. Then we define $z'(j) = s'(j')(j'')$. To see that this z' belongs to X' we show that if k, m of J are both $\leq (c^2j)$, then $(z'(k), z'(m)) \in U'^*(j)$; for if we choose an n that is $\leq k'$ and $\leq m'$, then $(s'(k'), s'(n))$ and $(s'(m'), s'(n))$ are both in $U'^*(c^2j)$, since they are respectively in $U'^*(c^2k)$ and $U'^*(c^2m)$ with $c^2k \leq k \leq c^2j$ and $c^2m \leq m \leq c^2j$. Hence there are nonnull initial sets B', B'' of $(J, \leq)$ such that $(s'(k')(t), s'(n)(t)) \in U^*(c^2j)$ for t in B' and $(s'(m')(t), s'(n)(t)) \in U^*(c^2j)$ for t in B''; if we then take the nonnull initial set $C = B' \cap B'' \cap \wedge(k'') \cap \wedge(m'')$, for a t from this C we would have that $U^*(c^2j)$ contains all the following pairs: $(z'(k), s'(k')(t))$, $(s'(k')(t), s'(n)(t))$, $(s'(n)(t), s'(m')(t))$, and $(s'(m')(t), z'(m))$. It follows that $(z'(k), z'(m)) \in U'^*(j)$.

Next we show that the sequence s' converges to z' in $(X', T(U'^*))$. If $m \leq j'$, we shall show that $(s'(m), z') \in U'^*(j)$ (for any given j of J). Given such an $m \leq j'$, we choose n in J such that $[k, h \leq n]$ would imply that $[(s'(m)(k), s'(m(h)) \in U^*(c^2j)]$, using the fact that $s'(m)$ is a Cauchy sequence of (X, U). Choosing a k from the nonnull initial set $\wedge(c^2j) \cap \wedge(n)$ and a $t \leq (m$ and $k')$, we have that $U'^*(c^2k)$ contains each of the pairs $(s'(k'), s'(t))$, $(s'(t), s'(m))$; hence there are nonnull initial sets C', C'' of $(J, \leq)$ such that for any p of C', $(s')(k')(p), s'(t)(p)) \in U^*(c^2j)$, and for a p of C'', $(s'(t)(p), s'(m)(p)) \in U^*(c^2j)$. Choosing a p from the nonnull initial set $C' \cap C'' \cap \wedge(k') \cap \wedge(n)$, we have, besides the two preceding relations, the pairs $(s'(k')(k''), s'(k')(p))$ and $(s'(m)(p), s'(m)(k))$ belonging to $U^*(c^2j)$. From these four we have then $(z'(k), s'(m)(k)) \in U^*(j)$ for all k in a certain nonnull initial set of $(J, \leq)$. Hence $(s'(m), z')$ is in $U'^*(j)$ for any $m \leq j'$. Thus s' converges to z' in $(X', T(U'^*))$. The semiuniform space $(X', U', (J, \leq), c)$ is thus complete. If E' is the equivalence on X' defined by U', $(E' =$

$\cap[U'^*(j)])$, the semiuniform quotient T_0-space $(X'/E', U'/E', (J, \leq), c)$ can also be shown to be complete like (X', U'). It is this quotient space that we define as the canonical completion $(X\#, U\#)$.

When the original semiuniformity is a quasiuniformity, then so are U' and $U\#$, so that the canonical completion is now a uniform space.

Any point s' of X' can be viewed as a limit in $(X', T(U'^*))$ of a Cauchy $(J, \leq)$-sequence $c(s')$ of the space (X', U'), namely, the sequence $c(s')$ defined by $(c(s'))(j)$ equals the constant sequence $[s'(j)]$ of (X, U). Similarly the typical point $E'(s')$ of $X\#$ (the E'-class containing an s' of X') can be viewed as a limit of a Cauchy $(J, \leq)$-sequence, the image of $c(s')$ by the canonical surjective map of X' on $X\# = X'/E'$. This last space, being a completely regular T_0-space, is T_2' also, hence a Hausdorff space. So limits of sequences are unique in this space, and any point of $X\#$ is the unique limit of a Cauchy $(J, \leq)$-sequence from $(X\#, U\#)$.

We can define the mapping h of X in $X\#$ by $h(x) = p'(x^c)$, which equals the image under the canonical surjection p' (of X' on $X\#$) of the constant $(J, \leq)$-sequence, of (X, U) whose values are all equal to x ($x^c(j) = x$ for all j). This h can be seen to be a uniform map of (X, U) in $(X\#, U\#)$: $(x, y) \in U(cj)$ implies that $(h(x), h(y))$ is in $U\#(j)$ for any j of J. If g: $(X, U) \to (Y, V, (K, \leq), c)$ is a uniform map and (Y, V) is complete and T_0, we define the map $g\#$: $X\# \to Y$ by $g\#(E'(s'))$ equals the unique limit in $(Y, T(V^*))$ of the Cauchy $(J, \leq)$-sequence $g(s')$ in (Y, V). Since Cauchy sequences s', t' of (X, U) for which $s'E't'$ give rise to images $g(s')$, $g(t')$ which are equivalent Cauchy sequences that have the same limit in the CR-, T_0-space $(Y, T(V^*)$, this definition is really unambiguous; it does not depend on the choice of the s' from the E'-class $E'(s')$. This $g\#$ can be seen to be a uniform map, with $g\# \cdot h = g$. Also, it can be shown that if $g'' \cdot h = g$ for a uniform map g'': $(X, U) \to (Y, V)$, then $g'' = g\#$. Thus the $g\#$ is unique. (See Krishnan [19], Theorem 1.)

If the semiuniformity is defined by a semimetric d on X, we can take the semiuniformity to be $(S, (N^*, \geq), c^*)$ as before. Then X' consists of Cauchy $(N^*, \geq)$-sequences s' from (X, S). We can then define a semimetric d' for X' by $d'(s', t') = \lim[d(s'(n), t'(n)]$, as n tends to infinity. This would give the semiuniformity (U') for X'; and the semiuniformity $U\#$ can be obtained from a semimetric $d\#$ on $X\#$ given by $d\#(s\#, t\#) = d'(s', t')$ for any choice of s', t' from $s\#$, $t\#$. That these do determine well-defined semimetrics that give the associated semiuniformities for X', $X\#$ can be proved. (This is an exercise for the reader!) □

25. Starting with the semimetric space (Q, d) with $d(x, y) = \max(0, y - x)$, show that the canonical completion can be identified (up to equivalence) with the semimetric space (R, d): the space of reals with d defined as above again.

26. By a *topological group* $(G, \cdot, ( )', 1, \mathbf{H})$ we mean a group $(G, \cdot, ( )', 1)$ together with a topology $\mathbf{H}$ on G such that the mappings $(x, y) \to (x \cdot y)$ of $G \times G$ in G and $x \to (x)'$ of G in G are both continuous, where $G \times G$ is taken with its product topology. A semiuniform group

$(G, \cdot, (\ )', 1, U, (J, \leq), c)$ is a group, with a semiuniformity on G, such that the foregoing maps are uniform maps (again taking $G \times G$ with its product semiuniformity), while for the map $(x) \to (x)'$ we consider it to be from (G, U^r) to (G, U). Verify that a semiuniform group determines a topological group when we use the topology defined by the semiuniformity. Moreover, the converse is also true for any topological group: there is a semiuniformity defining the topology that makes the group a semiuniform group too.

27. Given a semiuniform group, show that its canonical completion can be assigned a group structure under which it becomes a semiuniform group, with the uniform map h also a group homomorphism now.

28. In particular, show that the foregoing construction leads from the semimetrizable semiuniform group of rationals to the semimetrizable semiuniform group of the reals when we use semimetrics as in Exercise 25.

CHAPTER TWO

CATEGORIES, DEFINITIONS, AND EXAMPLES

2.1 CONCRETE AND GENERAL CATEGORIES

Two isomorphic algebras have essentially the same properties. So also a pair of isometric semimetric (or metric) spaces are practically equivalent, and homeomorphic spaces are indistinguishable, except in the names of the points. These special mappings, like isomorphisms, isometries, or homeomorphisms, are really special forms of homomorphisms, contraction maps, or continuous maps. Many structural properties can be formulated by using these mappings and their compositions. To have a common base for all these types of structures, we incorporate in the definition of a concrete category the common features that all these special examples share.

Definition 2.1.1 A *concrete category* $C = (O, U, \text{hom})$ consists of (i) a class of elements O called the *objects* of C; (ii) for each object A a set $U(\text{A})$, called the underlying set of A; and (iii) for each pair of objects, A, B, a set $\text{hom}_C(\text{A, B})$ of maps from $U(\text{A})$ to $U(\text{B})$, called the *morphisms* from A to B in C, these being subject to the following conditions:

(a) for each object A, $\text{hom}_C(\text{A, A})$ contains $1_{U(\text{A})}$;
(b) for objects A, B, C from C, when f, g are from $\text{hom}_c(\text{A, B})$ and $\text{hom}_c(\text{B, C})$, respectively, the composite map $g \cdot f$ belongs to $\text{hom}_c(\text{A, C})$.

We shall consider several examples. When dealing with different categories, it is convenient to tack on the name of the category and write $O(C)$, $U(C)$ for the O, U pertaining to the category C; we have already put in the C for the hom as a subscript. We also denote the family of all morphisms in C [the union of the $\hom_C(A, B)$ as A, B range over the class $O(C)$] by $M(C)$.

Example 1. The category S of sets: With $O(S)$ = [class of all sets], $U_S(A) = A$ for any A from $O(S)$, and $\hom_S(A, B)$ = [set of maps f: $A \to B$].

Example 2. Some algebras: *Gpd*: with $O(Gpd)$ = [class of groupoids $(A, \cdot)$]; $U(A, \cdot) = A$; $\hom_{Gpd.}((A, \cdot) \to (B, \cdot))$ = [maps f: $A \to B$, which are groupoid homomorphisms of $(A, \cdot)$ in $(B, \cdot)$]. Similarly, *Sgp* has semigroups for objects, underlying sets as above, and morphisms from $(A, \cdot)$ to $(B, \cdot)$ again are groupoid homomorphisms. *RSgp* and *ISgp* have regular semigroups and inverse semigroups, respectively, for objects and regular groupoid homomorphisms for morphisms. More generally, we can set *FDA* for a category whose objects are *FD*-algebras and whose morphisms are homomorphisms for all the operations in F. We could also consider a category (FDA, F') where F' is a subset of F; the objects of this category are still *FD*-algebras, but morphisms are maps that are homomorphisms for the operations from F' only, for example, a category whose objects are groups and whose morphisms are only semigroup homomorphisms between groups.

Example 3. Topological spaces: T, T_0, T_1', T_2', T_3, T_4, CR, and $T_2'COM$ are categories whose objects are, in order, arbitrary topological spaces, T_0-spaces, T_1'-spaces, T_2'-spaces, T_3-spaces, T_4-spaces, completely regular spaces, and T_2'- and compact spaces. The underlying set for any such space is the usual set base, and the morphisms are continuous maps between spaces in the various cases.

Example 4. Uniform spaces: *SU*, *QU*, and *U* are categories whose objects are the semiuniform spaces, quasiuniform spaces, and uniform spaces, respectively. In each case, the morphisms are uniformly continuous maps.

Example 5. Metric spaces: *SM*, *QM*, and *M* denote the categories with objects consisting of semimetric, quasimetric, or metric spaces. For each of these, a typical morphism f: $(X, d) \to (Y, d')$ is a set map of X in Y such that for all x, x' from X, $d'(f(x), f(x')) \leq d(x, y)$. These are called *contraction mappings*.

Example 6. The categories *PS*, *PTOP* of pointed sets and pointed spaces: $O(PS)$ = [(X, p): X a set, p in X]; $\hom_{PS}[(X, p), (Y, q)]$ = [f: $X \to Y$: $f(p) = q$]; $U(X, p) = X$. $O(PTOP)$ = [[$(X, G; p)$: $(X, G) \in O(TOP$, $p \in$

X]; $\hom_{PTOP}[(X, G; p), (Y, H; q)] = [f: (X, G) \to (Y, H)$ from $M(TOP)$, and $f(p) = q]$; $U[(X, G; p)] = X$.

Example 7. The category V of (real) vector spaces, and the linear transformations between pairs of these; or *BANS* of Banach spaces, and contraction mappings between pairs of these; or *R-Mod* of modules over a ring R with unit, and homomorphisms between pairs of these.

Definition 2.1.2 A category C consists of: a class O of C-objects, a class M of C-morphisms, a pair of mappings "dom" and "cod" of M in O (the images under these of an element f of M are called the domain of f and the codomain of f, respectively), and a mapping (or composition) " · " of $D = [(f, g): f, g$ from M with $\mathrm{dom}(f) = \mathrm{cod}(g)]$ in M:the image of the pair (f, g) from D under this mapping is usually denoted by $f \cdot g$. These are subject to the following axioms:

(i) When (f, g) is in D (or, as we may also express it: when $f \cdot g$ is defined) $\mathrm{dom}(f \cdot g) = \mathrm{dom}(g)$ and $\mathrm{cod}(f \cdot g) = \mathrm{cod}(f)$.
(ii) When (f, g) and (g, h) are both in D, $(f \cdot g) \cdot h = f \cdot (g \cdot h)$.
(iii) To each A of O is associated an identity morphism e of M such that $\mathrm{dom}(e) = A = \mathrm{cod}(e)$, where e of M is called an identity morphism when (e, e) is in D and $e \cdot f = f$ whenever (e, f) is in D, and $g \cdot e = g$ whenever (g, e) is in D.
(iv) For each pair A, B from O, $\hom_C(A, B) = [f: f$ in M with $\mathrm{dom}(f) = A, \mathrm{cod}(f) = B]$ is a set.

We note that for identity morphisms e, f from M, if $e \cdot f$ is defined then e must be the same as f, and thence it is easily seen that there is only one e that is an identity morphism and satisfies $\mathrm{dom}(e) = A = \mathrm{cod}(e)$, for any A of O. This unique e which exists by axiom (iii) is denoted by 1_A.

We also note that when $\hom_C(A, B)$ and $\hom_C(A', B')$ have a common element f then $A = \mathrm{dom}(f) = A'$ and $B = \mathrm{cod}(f) = B'$. Thus each f of M belongs to a unique set of the form $\hom_C(A, B)$, and M can be viewed as a disjoint union of these sets $[\hom_C(A, B)$: for all the distinct pairs (A, B) from $O]$.

The above remarks show that we can fully describe an abstract category C by specifying the class of C-objects, the sets $\hom_C(A, B)$ for pairs of objects (A, B), the identity morphisms 1_A corresponding to the objects A, and the mode of composing a pair (f, g) to get $f \cdot g$, when g is from $\hom_C(A, B)$ and f from $\hom_C(B, C)$ for some objects, A, B, C,with $f \cdot g$ being in $\hom_C(A, C)$. Then one has only to check the associativity of this composition and the characteristics of 1_A.

The following can be seen to be examples of general categories as described above.

Example 8. A concrete category $C = (O, U, \text{hom})$ as defined in 2.1.1 gives rise to (and is fully characterized by) an associated category C^* when we set O = the class of C^*-objects, $\text{hom}_{C^*}(A, B) = [(A, f, B)$: with f in $\text{hom}_C(A, B)]$, $1_A = (A, I_{U(A)}, A)$, ($I_{U(A)}$ denotes the identity map in the category of sets) and $(B, f, C) \cdot (A, g, B) = (A, f \cdot g, C)$.

Example 9. *ENS*: The objects of this category are arbitrary sets; for sets A, B, $\text{hom}_{ENS}(A, B) = [(A, \sigma, B)$: where σ is any relation from A to B, or any subset of $A \times B]$, $1_A = (A, 1_A, A)$, and $(B, \rho, C) \cdot (A, \sigma, B) = (A, \rho \circ \sigma, C)$.

Example 10. $(P, \leq)$: When $(P, \leq)$ is a any preordered set, we can treat it as a category with P for its class (or set) of objects, $\text{hom}_{(P,\leq)}(p, q)$ = either the single pair (p, q) when $p \leq q$, or the null set when $p \nleq q$, $1_p = (p, p)$, and $(q, r) \cdot (p, q) = (p, r)$ when $p \leq q \leq r$.

This last example can be characterized as a typical small, preordered category, according to the following definition.

Definition 2.1.3 A category C is called (i) *small* if the class of C-objects is a set, (ii) *discrete* if $\text{hom}_C(A, B)$ is null when $A \neq B$ and $\text{hom}_C(A, A) = [1_A]$, and (iii) *preordered* if, for any pair of objects A, B, $\text{hom}_C(A, B)$ has at most one element.

In the last type, if we set $A \leq B$ whenever $\text{hom}_C(A, B)$ is nonnull, this gives a preorder on the class of C-objects.

Since the axiom (iii) in the definition of a category C induces a one-to-one correspondence between the class of C-objects and a class of identity morphisms we can define a category in terms of the class of C-morphisms and this subclass of identity morphisms, without any explicit mention of objects. This is seen in

Lemma 2.1.1 *A category C given by (O, M, dom, cod, ·) determines and is fully determined by a triple (M, E, ·) consisting of a class M, a subclass E of M and a partial binary composition "·" defined in M, subject to the following conditions:*

(1) Each e of E is an identity element of (M, ·): that is, $e \cdot e$ is defined; whenever $e \cdot f$ is defined $e \cdot f = f$ and whenever $g \cdot e$ is defined $g \cdot e = g$.
(2) For each f in M, there exist e_1, e_2 in E such that $f \cdot e_1$ and $e_2 \cdot f$ are defined.

(3) *For a triple of elements f, g, h from M, we have*

> [$f \cdot g$ *and* $g \cdot h$ *are both defined*] $\Leftrightarrow$ [$f \cdot g$ *and* $(f \cdot g) \cdot h$ *are both defined*] $\Leftrightarrow$ [$g \cdot h$ *and* $f \cdot (g \cdot h)$ *are both defined*] $\Rightarrow$ [$f \cdot (g \cdot h) = (f \cdot g) \cdot h$].

(4) *For each pair of elements* e_1, e_2 *from E, the class* $hom(e_1, e_2)$ $=$ [f: f *in M*; $f \cdot e_1$ *and* $e_2 \cdot f$ *are both defined*] *is a set.*

PROOF. Starting from a category $C = (O, M, \text{dom}, \text{cod}, \cdot)$ as defined earlier, if we set $E = [1_A\text{: A in } O]$, it is easy to verify that the triple $(M, E, \cdot)$ satisfies the above four conditions.

Conversely, assuming a triple satisfying the four conditions, we set $C = (E, M, \text{dom}, \text{cod}, \cdot)$, where we set dom($f$) = [$e_1$ in E: $f \cdot e_1$ is defined] and cod(f) = [e_2 in E: $e_2 \cdot f$ is defined]. The condition (2) shows that such e_1, e_2 exist for any f. That these are uniquely determined by f can be shown, from properties of the identities e_1, e_2. Then we have only to verify that $f \cdot g$ is defined in M iff dom(f) = cod(g), and this is not hard, using the four given properties.

These two passages between the first form of the definition and this one in terms of the triple really are the reverse of each other and establish their essential equivalence. □

2.2 SUBCATEGORIES AND QUOTIENT CATEGORIES

Groups are special cases of semigroups and compact spaces are special types of topological spaces; these then give subcategories of larger ones.

Definition 2.2.1 A category C is called a *subcategory* of a category D if $O(C)$ is a subclass of $O(D)$, $M(C)$ is a subclass of $M(D)$, for any A, B from $O(C)$ hom_C(A, B) is a subset of hom_D(A, B), and for two morphisms f, g from $M(C)$ with dom f = cod g the products $f \cdot_C g$ and $f \cdot_D g$ of g followed by f are the same in both categories. If, further, it is also true that for any A, B from $O(C)$, hom_C(A, B) = hom_D(A, B), we say that C is a *full subcategory* of D.

From Examples 1–9 we have many such subcategories: *Sgp*, *RSgp*, *ISgp*, and *Gp* are all subcategories of *Gpd*. Of these, only the first is a full subcategory; *ISgp* is a full subcategory of *RSgp*; generally, $(F_1 D_1 A, F_1')$ is a subcategory of $(F_2 D_2 A, F_2')$ provided F_1 contains F_2, D_1 contains D_2, and F_1' contains F_2'; it is even a full subcategory if $F_1' = F_2'$. The space categories T_0, T_1', T_2', T_3, T_4, *CR*, $T_2'COM$ are all full subcategories of T, etc.

Given a category C, an equivalence relation E on the class $M(C)$ is called a *congruence* on C if (i) $[(f, g) \in E] \Rightarrow [\text{dom} f = \text{dom}\, g, \text{cod} f = \text{cod}\, g]$, and (ii) $[(f, g) \in E, (f', g') \in E, f \cdot f' \text{ is defined}] \Rightarrow [((f \cdot f'), (g \cdot g')) \in E]$.

Definition 2.2.2 Given a category C and a congruence E on C, we define the *quotient* category $C/E = D$ of C by E as follows: $O(D) = O(C)$; $M(D) =$ [the E-classes $E(f)$: f in M(C)] where $E(f) = [f'$ in $M(C)$: $(f, f') \in E]$; $\text{hom}_D(\text{A, B})$ (for A, B from $O(D) = O(C)) = [E(f); f$ in $\text{hom}_C(\text{A, B})]$; and [$F \cdot G$ is defined for F, G from $M(D)$] iff [there is an f in F and a g in G for which $f \cdot g$ is defined], and then $F \cdot G = E(f \cdot g)$; the choice of such a pair f, g from F, G does not affect the end result $E(f \cdot g)$.

That this does indeed satisfy all the conditions for a category D can be verified. We consider two important types of such quotients.

Example 11. $CREL^*$ is defined as follows: $O(CREL^*) = [(\text{X}, R)$: where X is a set and R is an equivalence relation on X]; $\text{hom}_{CREL^*}((\text{X}, R), (\text{Y}, S)) = [f$: a cofull relation from X to Y with $f \circ R \circ f^r$ contained in S]; the partial operation in $M(CREL^*)$ is relational multiplication, defined for f, g when $\text{dom} f = \text{cod}\, g$. On this category $CREL^*$ we can define a congruence E by $[(f, g) \in E]$ iff $[\text{dom} f = \text{dom}\, g$, $\text{cod} f = \text{cod}\, g$, and $f \circ R \circ g^r$ and $g \circ R \circ f^r$ are both contained in S]. It is not difficult to verify that E is indeed a congruence on this category $CREL^*$. The quotient category $CREL^*/E$ can then be formed. A special case of this construction is when we start with S^*, the subcategory of $CREL^*$ with objects that are the same as in $CREL^*$ but morphisms taken as mapping relations. E and the quotient S^*/E can again be defined.

Example 12. *HTOP* and *HPTOP*: On the category *TOP* a congruence E can be defined by setting (f, g) in E whenever f, g are continuous maps between the same pair of spaces (X, G), (Y, H) and there is a homotopy between f and g; similarly, we use homotopy which keeps the point p going to q, between pointed spaces (X, G; p) and (Y, H; q), to define a congruence E on *PTOP*. The resulting quotient categories, denoted by *HTOP* and *HPTOP*, are very important in algebraic topology.

2.3 PRODUCTS AND COPRODUCTS OF CATEGORIES

Given an indexed set of categories $[C_j$: j in $J]$, we associate with it two categories—$C = P[C_j]$, called the *product* of the family; and $D = U[C_j]$, called the *coproduct* (or disjoint union) of the family—as follows:

$$O(C) = [\text{maps } C\colon J \to U(O(C_j)) \text{ with } C(j) \text{ in } O(C_j) \text{ for each } j],$$

$$M(C) = [\text{maps } f\colon J \to U(M(C_j) \text{ with } f(j) \text{ in } M(C_j) \text{ for each } j],$$

$$\hom_C(A, B) = [f \text{ in } M(C) \text{ with } f(j) \text{ in } \hom C_j(A(j), B(j)) \text{ for each } j]$$

for any A, B from $O(C)$, and $f \cdot g = h$ for f, g, h in $M(C)$ iff $f(j) \cdot g(j) = h(j)$ for each j in J in the corresponding $M(C_j)$;

$$O(D) = U[O(C_j) \times (j)],$$

$$M(D) = U[M(C_j) \times (j)],$$

these being disjoint unions, $\hom_D((A, j), (B, k))$ is null if $j \neq k$, and $\hom_D((A, j), (B, j)) = [(f, j)\colon \text{for } f \text{ in } \hom_{C_j}(A, B)]$, noting that A, B are from $O(C_j)$. Then $(f, j) \cdot (g, k)$ is defined in $M(D)$ only when $j = k$ and $f \cdot g$ is defined in $M(C_j)$, and then $(f, j) \cdot (g, j) = (f \cdot g, j)$.

When the indexing set J is the finite set $(1, 2, \ldots, n)$, we also write $C_1 \times C_2 \times \cdots \times C_n$ for the product and $C_1 \cup C_2 \cup \cdots \cup C_n$ for the coproduct. Again, for any indexing set J, if the C_j are all the same as C_0, we write $C_0{}^J$ and $J \cdot C_0$ for the product and coproduct.

2.4 THE DUAL CATEGORY AND DUALITY OF PROPERTIES

To each category C we can associate a *dual category* C^{op} by setting $O(C^{\text{op}}) = O(C)$, $M(C^{\text{op}}) = M(C)$, $\hom_{C^{\text{op}}}(A, B) = \hom_C(B, A)$ for any A, B from $O(C) = O(C^{\text{op}})$; and for f, g, h from $M(C) = M(C^{\text{op}})$, $f \cdot g = h$ in $M(C)$ iff $g \cdot f = h$ in $M(C^{\text{op}})$. It is clear that the dual of C^{op} is C itself.

Usually we visualize a category as a class of points for the objects, and a class of arrows for the morphisms, each arrow going from the point that is its domain to the point that is its codomain. For finite categories these diagrams can be drawn on paper. The dual of a category is then pictured as one with all the names for objects as well as morphisms unchanged and only the direction of each arrow reversed.

By a categorical property P we mean one such that for any category it makes sense to say that P is either "true" or "not true" in the category. Given such a property P, we can specify a dual property P^{op} by saying that P^{op} is true in a category C iff P is true in the dual category C^{op}. If $[Q_j]$ is a family of such properties that contains with each Q its dual Q^{op} also, then it is clear that if a property P can be proved for each category satisfying the properties $[Q_j]$, then P^{op} can also be proved for each category satisfying the properties $[Q_j]$. This *duality* principle is often utilized in the following in proving dual properties.

2.5 ARROW CATEGORY AND COMMA CATEGORIES OVER A CATEGORY

We pictured a category by arrows joining points. Such a diagram, or part of such a diagram, is called commutative if each chain of arrows connecting the same pair of end points gives rise to a composite morphism between the objects represented by the points that is unaltered by taking different such chains (in the part of the diagram in question). For instance the diagrams in Figure 2.1 give commutative diagrams iff $f \cdot g = h$ for the arrows in the triangle and $f \cdot g = h \cdot k$ for the arrows in the square. Moreover, there must be a diagonal morphism m if all morphisms (other than identities) are to be marked by arrows in the diagram. Often we do not require this, and let composites of morphisms be seen as a chain of successive arrows.

Given a category C, we can form a category $A(C)$ whose objects are the arrows, or morphisms, of C. We call this the *arrow category* over C; it is specified by $O(A(C)) = M(C)$, $M(A(C)) =$ [quartets (h, f, g, h') of elements from $M(C)$ satisfying $h \cdot f = g \cdot h'$], $\hom_{A(C)}(f, g) =$ [(h, f, g, h'): with h, h' from $M(C)$ that belong to $M(A(C))$, that is, satisfy $h \cdot f = g \cdot h'$]; the product of (h, f, g, h') and (k, s, t, k'), that is, $(k, s, t, k') \cdot (h, f, g, h')$, is defined in $A(C)$ iff $\mathrm{dom}(k, s, t, k') = s = g = \mathrm{cod}(h, f, g, h')$, and then it is $(k \cdot h, f, t, k' \cdot h')$. (See Figure 2.2.)

To define the *comma categories* we start with a category C and an object A chosen from it. Then we have two associated comma categories, $(\mathrm{A} \rightarrow C)$ and $(C \rightarrow \mathrm{A})$, called the *comma category of* A *over* C and the *comma category of* C *over* A, respectively. The objects of the first are morphisms of C with domain A; those of the latter are morphisms of C with codomain A. A typical morphism of the first category is a triple (f, g, f') from $M(C)$ such that $g \cdot f = f'$ and $\mathrm{dom}\, f = \mathrm{A}$, while for the second a typical morphism is a triple (f, g, f') from $M(C)$ with $f' \cdot g = f$ and $\mathrm{cod}(f) = \mathrm{A}$. The composition of two such triples (f, g, f') followed

Figure 2.1

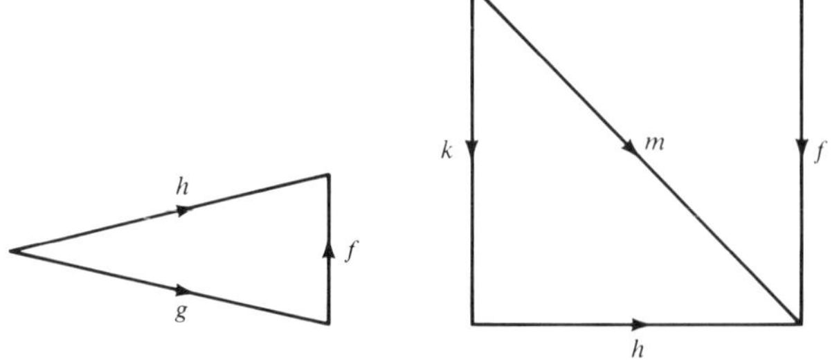

h
f
g
g
k
m
f
h

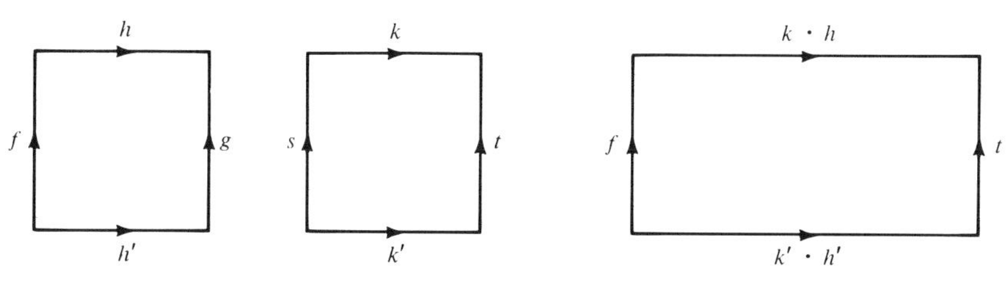

Figure 2.2

by (h, k, h') is defined (in either category) iff $f' = h$ and $k \cdot g$ is defined in $M(C)$, and then the composite is $(f, k \cdot g, h')$. (See Figure 2.3.)

EXERCISES AND REMARKS ON CHAPTER 2

1. Using the various types of topological spaces (such as T_0-, T_3-, CR-spaces) as objects but taking for morphisms either the "open maps" or the "closed maps," we get whole new series of categories of spaces, which we denote by (T_0, O), (CR, Cl), etc. (A map f of a space (X, G) in another (Y, H) is called an open (a closed) map if it takes open (closed) sets of (X, G) to open (closed) sets in (Y, H)). Show that $(T_2'COM)$ is a subcategory of (T_2', Cl). Is it a full subcategory?

2. For the semiuniform, quasiuniform, or uniform spaces as objects we may take the morphisms to be maps between two such spaces that are merely continuous under the induced topologies, not uniformly continuous. Show that (SU) is a subcategory of (SU, Con), the latter having continuous maps for morphisms; show also that it is not a full subcategory.

3. For semimetric, quasimetric, or metric spaces as objects we may take the morphisms to be merely uniformly continuous maps, not contraction maps; such a category, say, (SM, Ucon), contains (SM) as a subcategory that is not full.

4. For the finite categories A, B given by Figure 2.4, draw diagrams representing $A \times B$ and $A \cup B$. How many objects and morphisms do each of these new categories have?

Figure 2.3

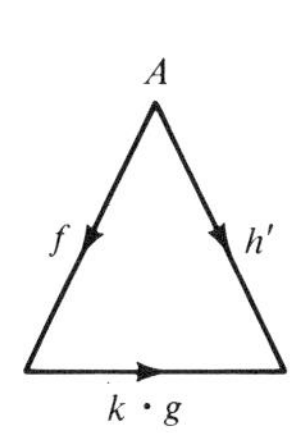

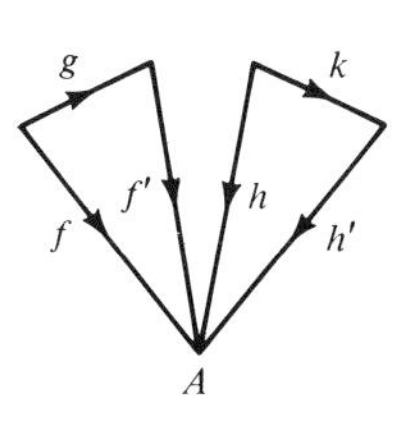

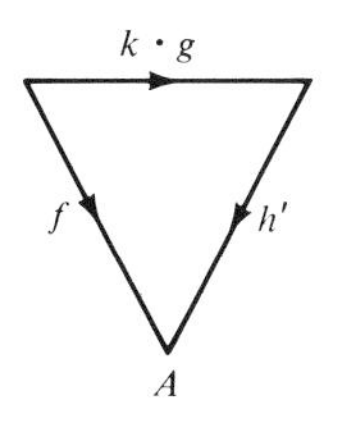

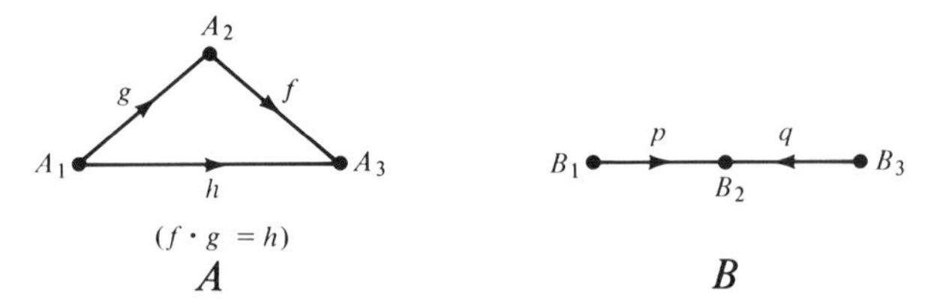

Figure 2.4

5. If $A_1, A_2, \ldots, A_n$ are all the same as the category A, check that their coproduct $A_1 \cup \cdots \cup A_n$ looks just like the category $A \times B$ if B denotes the discrete category of n elements $(1, 2, \ldots, n)$.

6. Treating a preordered set $(P, \leq)$ as a small preordered category, check that its dual looks like $(P, \geq)$ (the dual preordered set).

7. State the duals of the following properties:
(i) the morphism f has a left inverse g (that is, $g \cdot f = 1_{\mathrm{dom}(f)}$);
(ii) the morphism f is left-cancelable (that is, $f \cdot g = f \cdot h$ implies $g = h$).

8. Show that the category C has just one object iff $(M(C), \cdot)$ is a semigroup with identity.

CHAPTER THREE

DISTINGUISHED MORPHISMS AND OBJECTS

3.1 DISTINGUISHED MORPHISMS

The injective and surjective maps have special properties in the category $S\#$ of nonnull sets and maps. We start with these in

Lemma 3.1.1 *A morphism f: A $\to$ B in the category $S\#$ is injective iff it satisfies either of the following two conditions: (a) f is left-cancelable, that is, $(f \cdot k = f \cdot h)$ implies $(k = h)$; (b) f has a left-inverse g, that is, $g \cdot f = 1_{\mathrm{dom}(f)}$. Dually, f is surjective iff it satisfies either of these conditions: (a′) f is right-cancelable; (b′) f has a right-inverse g.*

PROOF. If f is injective as a mapping from the nonnull set A to B, for each b in B either no a exists in A with $f(a) = b$ or there is just one a in A with $f(a) = b$; set $g(b)$ to equal this a if $f(a) = b$, and set $g(b) = a_0$, some chosen element of A, if no $f(a) = b$. Then g is a map of B in A with $g \cdot f = 1_{\mathrm{dom}(f)} (= 1_A)$. If next f has a left-inverse g, with $g \cdot f = 1_A$, then $(f \cdot h = f \cdot k)$ implies $g \cdot (f \cdot h) = g \cdot (f \cdot k)$ or $(g \cdot f) \cdot h = (g \cdot f) \cdot k$ or $h = k$. Finally, if g is left-cancelable, f cannot but be injective: for if it were not, there must be elements a, a' in A that are distinct for which $f(a) = f(a')$; defining $h = 1_A$ and k: A $\to$ A as the map that interchanges a with a' and leaves all other elements of A unchanged, we have an $h \neq k$ with $g \cdot h = g \cdot k$.

The case of surjective maps is treated by a dual reasoning. □

This lemma suggests the following generalized notions in any category.

Definition 3.1.1 A morphism f: A $\rightarrow$ B in a category C is called

(i) a *monomorphism* if f is left-cancelable: that is, $[f \cdot k = f \cdot h] \Rightarrow [k = h]$;
(i′) an *epimorphism* if f is right-cancelable;
(ii) a *bimorphism* if f is left- and right-cancelable;
(iii) a *section* if f has a left-inverse g such that $g \cdot f = 1_A$;
(iii′) a *retraction* if f has a right-inverse g;
(iv) an *isomorphism* if f has a left- and right-inverse g.

Since $g \cdot f = 1_A$ implies that $[f \cdot k = f \cdot h]$ would give $k = (g \cdot f) \cdot k = g \cdot (f \cdot k) = g \cdot (f \cdot h) = (g \cdot f) \cdot h = h$, any section is a monomorphism; similarly, any retraction is an epimorphism. An isomorphism, being both a section and a retraction, is thus also a bimorphism. The converse may not always be true. We call the category C a *balanced category* if every bimorphism in C is an isomorphism.

When $C = (O, U, \text{hom})$ is a concrete category, it is easily seen that a morphism f from hom(A, B) is a monomorphism if it is an injective map from S and it is an epimorphism if it is a surjective map from S. In most of the familiar concrete categories, the monomorphisms are always injective maps; for many topological categories, the epimorphisms are surjective maps. But in (T_0, T_2'), the category of Hausdorff spaces (as these are called) and continuous maps between such, an epimorphism f from (X, G) to (Y, H) can be shown to be any continuous map of the first space in the second for which the image set $f(X)$ is dense in Y (that is, the closure of $f(X) = Y$). (See the exercises at the end of this chapter.) If there is a bijective map f: $U(\text{A}) \rightarrow U(\text{B})$ for objects A, B from the category C, f, f^r are in $M(C)$ iff f is an isomorphism. The objects A, B differ only in the way the elements of the underlying sets $U(\text{A})$, $U(\text{B})$ are named! They both have essentially the same structure, and generally we do not distinguish between such isomorphic objects. When we have a concrete category like Gp or TOP, a subgroup or subspace is defined by using the inclusion map of the underlying set of one object in that of another. This map is a monomorphism in the category. But in general categories we have no way of choosing such a distinguished monomorphism. Thus our definition of subobject is based just on the notion of a monomorphism, and dually our definition of a quotient object on the notion of an epimorphism.

3.2 DISTINGUISHED OBJECTS

Definition 3.2.1 Given an object B of a category C, we call the pair (A, m) a *subobject* of B if A is an object of C and m a monomorphism m: A $\rightarrow$ B; dually, a pair (e, A) where A is an object of C and e is an epimorphism e: B $\rightarrow$ A is called a *quotient object* of B.

To compare different subobjects or quotient objects of the same object, we begin by observing that, given B from $O(C)$, in the classes of objects from the categories $(C \to \mathrm{B})$ and $(\mathrm{B} \to C)$ we can define preorder relations thus: for objects (A, f), (A', f') from $(C \to \mathrm{B})$ [we write (A, f) for an object of $(C \to \mathrm{B})$ when f; $\mathrm{A} \to \mathrm{B}$ is a morphism of C], $(\mathrm{A}, f) \leq (\mathrm{A}', f')$ iff there is a morphism (f, g, f') of $(C \to \mathrm{B})$ from the first to the second. That is, $f' \cdot g = f$! Similarly, for objects (f, A), (f', A') from $(\mathrm{B} \to C)$ we set $(f, \mathrm{A}) \leq (f', \mathrm{A}')$ iff there is a morphism (f, g, f') from the first to the second in $(\mathrm{B} \to C)$, so that $g \cdot f = f'$. If we restrict ourselves to the subclass of objects f from $(C \to \mathrm{B})$ that are monomorphisms from C, then for this subclass the restriction of this preorder is almost an order: for if each of (A, f) and (A', f') is less than or equal to the other and f, f' are monomorphisms of C, there exist g, g' such that $f \cdot g' = f'$ and $f' \cdot g = f$; so $f \cdot (g' \cdot g) = (f \cdot g') \cdot g = f' \cdot g = f = f \cdot 1_{\mathrm{A}}$; f being a monomorphism, $g' \cdot g = 1_{\mathrm{A}}$ follows. Similarly we see that $g \cdot g' = 1_{\mathrm{A}'}$; thus there is an isomorphism g with $f' \cdot g = f$. We write $(\mathrm{A}, f) \equiv (\mathrm{A}', f')$ when there is an isomorphism g with $f' \cdot g = f$, and call the subobjects (A, f) and (A', f') *equivalent subobjects*. Similarly, it can be seen that when each of two quotient objects (f, A), (f', A') of B is less than or equal to the other in $(\mathrm{B} \to C)$, then there is an isomorphism g with $g \cdot f = f'$; we call these *equivalent quotient objects* and write $(f, \mathrm{A}) \equiv (f', \mathrm{A}')$. These give equivalence relations in the classes of subobjects and quotient objects, respectively, of any object B. A subclass of the class of subobjects of B that contains at least one subobject from each equivalence class of subobjects is called a *representative subfamily* of subobjects of B. Similarly, a representative family of quotient objects of B would contain at least one quotient object from each equivalence class of them. The cases when such a representative family can be found that is a *set* are important; we define a category C to be *well-powered* (dually, *cowell-powered*) if there is a set of representative subobjects (a set of representative quotient objects) for each object B from C. It can be seen that a concrete category C is always well-powered when its monomorphisms are necessarily injective maps, and it is cowell-powered if the epimorphisms are necessarily surjective maps. These cover many of our earlier examples. (See the exercises at the end of this chapter.)

In S as well as in *TOP* there is an object with a null set as its set, and from this object there is only one morphism to any other object; similarly, there is an object with a one-element set as base such that from any object there is just one morphism to this object. This suggests the

Definition 3.2.2 An object A of the category C is called an *initial object* of C if for each object B of C $\hom_C(\mathrm{A}, \mathrm{B})$ contains exactly one morphism; dually, an object A of C is called a *final object* of C if for each object B from C $\hom_C(\mathrm{B}, \mathrm{A})$ contains exactly one morphism.

Evidently an initial or final object is unique up to an isomorphism if it exists, and any isomorph of an initial or final object is also such.

Definition 3.2.3 The object A of C is called a *zero object* of C if it is both an initial object and a final object of C.

Since S and TOP each has an initial object different from a final object, these categories cannot have a zero object at all. But Gr and $TOPGr$ are categories with a zero object based on a one-element set.

3.3 EQUALIZERS AND COEQUALIZERS

For a nonnull set $[f_j: j \text{ in } J]$ of morphisms of C all having the same domain and codomain, say all from $\hom_C(A, B)$, we call a pair (D, h) (or just h) an *equalizer for the family* (i) if h is from $\hom_C(D, A)$ and $f_j \cdot h = f_k \cdot h$ for all j, k from J; and (ii) if h' from $\hom_C(D', A)$ satisfies $f_j \cdot h' = f_k \cdot h'$ for all j, k from J, then there is a unique morphism g in $\hom_C(D', D)$ such that $h \cdot g = h'$. An equivalent formulation of this notion would be to describe (D, h) as a final object in the full subcategory of the comma category $(C \to A)$ with objects (D, h) that satisfy the condition $[f_j \cdot h = f_k \cdot h$ for all j, k from $J]$.

The most important case is when there are two morphisms (f, f') in the family. An equalizer of a pair of morphisms $(f, f') \in \hom_C(A, B)$ is called a *regular monomorphism*. The name would be justified when we show in the next lemma that such an equalizer h must be a monomorphism.

Dual notions are that of a *coequalizer* for a family of morphisms and that of a *regular epimorphism*, which is a coequalizer of a pair of morphisms (with common domain and codomain).

Lemma 3.3.1 *(i) An equalizer of any family of morphisms with common domain and codomain is a monomorphism.*

(ii) A section h from C is a regular monomorphism.

(iii) If h is a regular monomorphism and an epimorphism, then it is an isomorphism.

(iv) If h is a section and an epimorphism, then it is an isomorphism.

(v) The duals of all four of these results are true.

PROOF. (i) If h is an equalizer of the family $[f_j: j \text{ in } J]$ of morphisms from $\hom_C(A, B)$, and $h \cdot r = h \cdot s$ for some morphisms r, s from C, then since $f_j \cdot h = f_k \cdot h$ for all j, k from J, $f_j \cdot h' = f_k \cdot h'$ for all j, k of J when we set $h' = h \cdot r = h \cdot s$; then, from the definition of the equalizer there must be a unique g such that $h \cdot g = h'$; since both r and s qualify for being such a g, it follows that $r = s$. But this just proves that h is a monomorphism.

(ii) When h of $\hom_C(B, A)$ is a section in C it has a left-inverse k in $\hom_C(A, B)$ with $k \cdot h = 1_B$; now $h \cdot k$ and 1_A are both morphisms from $\hom_C(A, A)$; we claim that h is an equalizer for this pair, hence a regular monomorphism. First, $(h \cdot k) \cdot h = h \cdot (k \cdot h) = h \cdot 1_B = h = 1_A \cdot h$; whereas if for some h' we had $(h \cdot k) \cdot h' = 1_A \cdot h'$, then $h \cdot (k \cdot h') = h'$, so that there is a $g = k \cdot h'$ with $h \cdot g = h'$. This g is unique, for if $h \cdot g' = h'$, then $k \cdot h \cdot g' = k \cdot h'$ or $1_B \cdot g' = k \cdot h'$ or $g' = k \cdot h' = g$.

(iii) Suppose that h is an epimorphism and a regular monomorphism, therefore an equalizer for a pair $[f_1, f_2]$ of morphisms. From $f_1 \cdot h = f_2 \cdot h$, h being right-cancelable, it follows that $f_1 = f_2$; but then for the pair of equal morphisms (f_1, f_2) clearly an equalizer is $1_{\mathrm{dom}(f_i)} = 1_{\mathrm{cod}(h)}$; since any two equalizers are equal in the sense that each is a multiple of the other by an isomorphism on the right, h must be $1_{\mathrm{cod}(h)} \cdot j$ for an isomorphism j, which means $h = j =$ an isomorphism.

(iv) This follows from (iii), since a section is a regular monomorphism by (ii).

(v) Duals follows by the basic duality of the axioms defining the category. □

3.4 CONSTANT MORPHISMS AND POINTED CATEGORIES

In the category of sets S we call a mapping $f\colon A \to B$ a constant map if $f(a) = f(a')$ for all a, a' from A. In this case (and only then), it is easy to see that for any set D and any two maps g, h from D to A, $f \cdot g = f \cdot h$. This is the motivation for the definition that follows:

Definition 3.4.1 A morphism f from $\hom_C(A, B)$ in the category C is called a *constant morphism* if $[\forall D \in O(C)$ and $\forall g, h$ from $\hom_C(D, A)]$: $[f \cdot g = f \cdot h]$. Dually f is a *coconstant morphism* if $[\forall D \in O(C)$ and $\forall g, h$ from $\hom_C(B, D)]$: $[g \cdot f = h \cdot f]$. When f is both a constant morphism and a coconstant morphism it is called a *zero morphism*.

Lemma 3.4.1 *(a) When $\hom_C(B, A)$ is nonnull, [$\hom_C(A, B)$ contains a constant morphism and a coconstant morphism] $\Leftrightarrow$ [$\hom_C(A, B)$ contains a unique zero morphism] $\Leftrightarrow$ [$\hom_C(A, B)$ contains a zero morphism].*

(b) When C has a final object T, and $f \in \hom_C(A, B)$ factors through T [or $f = h \cdot g$ for a g in $\hom_C(A, T)$ and an h in $\hom_C(T, B)$], then f is a constant morphism; conversely, if f is a constant morphism in $\hom_C(A, B)$ and $\hom_C(T, A)$ is nonnull for the final object T, then f factors through T.

PROOF. (a) The point in (a) is to prove that when $\hom_C(B, A)$ is nonnull, [f is a constant morphism and g a coconstant morphism both in

$\hom_C(A, B)$] implies that [$f = g$]; for let k be any member of the nonnull set $\hom_C(B, A)$.

[$k \cdot g$ and 1_A are both in $\hom_C(A, A)$ and f is a constant morphism in $\hom_C(A, B)$] imply that $f \cdot (k \cdot g) = f \cdot 1_A = f$; (i)

[$f \cdot k$ and 1_B both are in $\hom_C(B, B)$ and g is a coconstant morphism] imply that $(f \cdot k) \cdot g = 1_B \cdot g = g$. (ii)

But in C we have the associativity $f \cdot (k \cdot g) = (f \cdot k) \cdot g$; hence (i) and (ii) together give $f = g$. From this result it follows that when $\hom_C(B, A)$ is nonnull and $\hom_C(A, B)$ contains both a constant morphism f and a coconstant morphism g, then $f = g$ is *a* zero morphism in $\hom_C(A, B)$; and second, if $\hom_C(A, B)$ contains two zero morphisms f, g, they are equal (because we can take one to be a constant and the other to be a coconstant morphism). Finally if $\hom_C(A, B)$ contains a (unique) zero morphism it certainly contains a constant and a coconstant morphism (namely, the zero morphism). These remarks prove part (a).

(b) If T is a final object of C and f of $\hom_C(A, B)$ has a resolution $f = h \cdot g$ with g in $\hom_C(A, T)$ and h in $\hom_C(T, B)$, for any object D of C and any pair of morphisms r, s from $\hom_C(D, A)$ we have $g \cdot r = g \cdot s$ is the unique morphism in $\hom_C(D, T)$. Hence $f \cdot r = h \cdot g \cdot r = h \cdot g \cdot s = f \cdot s$; thus f is a constant morphism. Conversely, if f is a constant morphism in $\hom_C(A, B)$, and we assume that $\hom_C(T, A)$ is nonnull, let k be a morphism from this nonnull set and v be the unique morphism in $\hom_C(A, T)$. Then, since $k \cdot v$ and 1_A are both in $\hom_C(A, A)$ and f is a constant morphism, we have $f \cdot (k \cdot v) = f \cdot 1_A = f$, or $f = (f \cdot k) \cdot v$, which is a factorization of f through T.

For example, in the categories S and TOP we have an initial object that is a null set (with obvious topology) and a final object (with a one-element set); the morphisms with the initial object as domain are zero morphisms; those with a final object as codomain are constant morphisms but not coconstant ones. In the categories of groups or rings we have a zero object (with a one-element set) and so a zero morphism f in each hom(A, B) [the composition $g \cdot h$ of the unique morphisms g, h from hom(O, B) and hom(A, O), if O is a zero object]. □

This last type suggests the following definition and study of "pointed categories."

Definition 3.4.2 A category C is called a *pointed category* if for each pair (A, B) from $O(C)$, $\hom_C(A, B)$ contains a zero morphism.

Theorem 3.4.1 *(a) In the category C, when the products $f \cdot g$ and $h \cdot f$ are defined, both these are constant (or coconstant, or zero) morphisms iff f is such a morphism.*

(b) The following are necessary and sufficient conditions for C to be a pointed category (they are both equivalent):

(i) for all A, B *from $O(C)$, hom_C(*A, B*) contains a constant morphism and a coconstant morphism;*

(ii) for all A, B *from $O(C)$, hom_C(*A, B*) contains a unique constant morphism (dually it contains a unique coconstant morphism).*

(c) If the category C has a zero object or is a full subcategory of a pointed category D, then C is a pointed category.

(d) Any pointed category can be exhibited as a full subcategory of a category with a zero object.

PROOF. (a) This result is useful in proofs of the later ones, and so is included here. If f, g, h are from $\hom_C$(A, B), $\hom_C$(D, A), and $\hom_C$(B, E) for A, B, D, E from $O(C)$, so that the products $f \cdot g$ and $h \cdot f$ are defined, and if f is a constant morphism, then for r, s from a $\hom_C$(F, D), since $g \cdot r$ and $g \cdot s$ are in $\hom_C$(F, A), $f \cdot (g \cdot r) = f \cdot (g \cdot s)$, so $(f \cdot g) \cdot r = (f \cdot g) \cdot s$. This proves that $f \cdot g$ is a constant morphism; if now r, s are from a $\hom_C$(G, A), then $f \cdot r = f \cdot s$ gives $(h \cdot f) \cdot r = h \cdot (f \cdot r) = h \cdot (f \cdot s) = (h \cdot f) \cdot s$, proving that $h \cdot f$ is also a constant morphism. The proofs are similar for f a coconstant morphism, and then the result follows for a zero morphism.

(b) That condition (i) is necessary and sufficient for C to be pointed follows from part (a) of Lemma 3.4.1, since when (i) is true surely $\hom_C$(B, A) is nonnull (for any B,A from $O(C)$). Assuming (i), we get (ii), since again $\hom_C$(B, A) is nonnull, and for any two constant morphisms f, f' in $\hom_C$(A, B) each of them must be equal to any coconstant morphism g in the same hom set, as shown in the proof of part (a), Lemma 3.4.1. Conversely, if (ii) is assumed, and f is the unique constant morphism in $\hom_C$(A, B), for any morphisms r, s from a $\hom_C$(B, D) we know, by part (a), that $r \cdot f$ and $s \cdot f$ are both constant morphisms in $\hom_C$(A, D). But there is only one constant morphism in $\hom_C$(A, D); hence $r \cdot f = s \cdot f$; thus f is also a coconstant morphism in $\hom_C$(A, B), proving that (i) is true.

(c) When O is a zero object in C, for each object A of C there is a unique morphism (OA) in $\hom_C$(O, A) and a unique morphism (AO) in $\hom_C$(A, O); thus for any A, B from $O(C)$, the morphism (OB) $\cdot$ (AO) factors through the zero object O, hence must be both a constant and coconstant morphism, by part (b) of Lemma 3.4.1. Thus C is a pointed category. When C is a full subcategory of a pointed category D, for objects A, B from C any zero morphism of D from $\hom_D$(A, B) is also a zero morphism in C from $\hom_C$(A, B). Hence C is also a pointed category.

(d) Starting with pointed category C, we know that for A, B from $O(C)$ there is a unique zero morphism in $\hom_C$(A, B), which we denote by (AB);

we now define an inclusive category D by setting $O(D) = O(C) \mathring{\cup} (\mathrm{O})$ where O is the symbol for a new object not in $O(C)$; $M(D)$ is the disjoint union of $M(C)$, 1_{O}, and two families [(AO): for A in $O(C)$] and [(OA): for A in $O(C)$]; and the composition of morphisms is governed by the requirements; $f \cdot g$ in $M(D)$ is the same as in $M(C)$ when f, g are both from $M(C)$; $1_{\mathrm{O}} \cdot 1_{\mathrm{O}} = 1_{\mathrm{O}}$ ($\mathrm{dom}(1_{\mathrm{O}}) = \mathrm{cod}(1_{\mathrm{O}}) = \mathrm{O}$); dom(AO) = A, cod(AO) = O, dom(OA) = O, cod(OA) = A; (OB) · (AO) = (AB); (AO) · (OA) $= 1_{\mathrm{O}}$; $f \cdot$ (OA) = (OB) for any f in $\mathrm{hom}_C(\mathrm{A}, \mathrm{B})$; (BO) $\cdot f$ = (AO) for any f in $\mathrm{hom}_C(\mathrm{A}, \mathrm{B})$. It is not hard to see that the identity morphisms in C and the zero morphisms in C continue to be identity morphisms and zero morphisms in D; that the associative law is true for triples of morphisms in D for which the law makes sense; and that O is a zero object in D, and C is the full subcategory of D that has for objects all those of D other than O. □

For example, if we consider the full subcategory of *Gp* whose objects are infinite groups, then this is a pointed category without a zero object; our last construction would add just one trivial one-element group to provide the missing zero object to this.

3.5 SEPARATORS AND COSEPARATORS

Definition 3.5.1 An object A from $O(C)$ is called a *separator* for C if for any B, D from $O(C)$ and any f, g from $\mathrm{hom}_C(\mathrm{B}, \mathrm{D})$ with $f \neq g$ there is an h in $\mathrm{hom}_C(\mathrm{A}, \mathrm{B})$ such that $f \cdot h \neq g \cdot h$. Dually a *coseparator* A is defined as an object of C such that for any f, g from $\mathrm{hom}_C(\mathrm{B}, \mathrm{D})$ with $f \neq g$ there is an h in $\mathrm{hom}_C(\mathrm{D}, \mathrm{A})$ such that $h \cdot f \neq h \cdot g$.

It is easy to see that if hom(A′, A) contains an epimorphism and A is a separator for C, so is A′; and dually. In the category S a one-element set is surely a separator, and hence also any nonnull set, by the last remark; and a two-element set is a coseparator, hence also any set containing at least two elements. In *TOP* any nonnull space is a separator, and a space with a nontrivial indiscrete subspace is a coseparator. In *Gp* and *AbGp* the additive group $(Z, +)$ is a separator.

EXERCISES AND REMARKS ON CHAPTER 3

1. Verify that composites of the following types of morphisms in a category also belong to the same type: mono-, epi-, or isomorphisms; sections; and retractions.

2. *The Algebras*

   (i) For any F, D the category *FDA* is well-powered and cowell-powered.
   (ii) In *FDA* a morphism f: $(X, F) \to (Y, F)$ is a monomorphism iff the un-

derlying set map $|f|$: $X \to Y$ is injective. [If $|f|$ is injective, it is clear that f would be a monomorphism. If $|f|$ were not injective, so $|f|(x) = |f|(x')$ for elements $x \neq x'$ from X, we could find morphisms g, h from the free *FD*-algebra $FD(a)$ over a one-element set (a), with $g(a) = x$, $h(a) = x'$, so that $f \cdot g = f \cdot h$ though $g \neq h$ This means that f is not a monomorphism.]

(iii) In *AbGp* and *R-Mod* a morphism f: $(X, +) \to (Y, +)$ is an epimorphism iff the set map $|f|$: $X \to Y$ is surjective. [Again, if $|f|$ is surjective, it is easy to see that f is epi-; if f is epi-, consider the two morphisms g, h of $(Y, +)$ in $(Y, +)/E$, where E is the congruence on $(Y, +)$ defined by the submodule $f(X)$, with g as the canonical map of $(Y, +)$ on $(Y, +)/E$ and h as the zero map (mapping all of Y on the zero of $(Y, +)/E$). Then, since $g \cdot f = h \cdot f$ (equals the zero map) and f is epi-, $g = h$ follows. Then it is clear that $|f|(X)$ must be equal to Y, or $|f|$ is surjective. It follows that *AbGp* and *R-Mod* are balanced categories.

(iv) In *Sgp* (or *Rn*, the category of rings), the inclusion map j from Z to Q is not surjective; but we can show that it gives an epimorphism j: $(Z, \cdot) \to (Q, \cdot)$ in *Sgp* (and in *Rn*). [For if g, h are morphisms from $(Q, \cdot)$ to an $(S, \cdot)$ such that $g \cdot j = h \cdot j$, then for a typical element (m/n) of Q we have $g(m/n) = g(m) \cdot g(1/n) \cdot h(1) = h(m) \cdot g(1/n) \cdot h(n) \cdot h(1/n) = h(m) \cdot g(1/n) \cdot g(n) \cdot h(1/n) = h(m) \cdot g(1) \cdot h(1/n) = h(m) \cdot h(1/n) = h(m/n)$.]

3. *The Spaces*

(i) A morphism f: $(X, G) \to (Y, H)$ between spaces in any category of topological spaces is a monomorphism (an epimorphism) if the set map f: $X \to Y$ is injective (surjective).

(ii) In the categories T, T_0, T_1', T_2', T_3, CR, and T_4, each of which contains a one-point space, when f is a monomorphism, the underlying set map is injective. (Otherwise we would have distinct morphisms g, h from the one-element space to the domain space for which $f \cdot g = f \cdot h$.)

(iii) In all the categories listed in (ii) except T_0, there is a two-point indiscrete space (with just two open sets). In these, when f is an epimorphism, then f must be surjective. [Otherwise we can find distinct morphisms g, h from the codomain space to the two-point indiscrete space, g, h differing at a point of the codomain outside the set $f(X)$, with $g \cdot f = h \cdot f$.]

(iv) In the *Hausdorff space category* (T_0 and T_2') it is known that a directed sequence can have at most one limit. A continuous map f: $(X, G) \to (Y, H)$ between two such spaces would be an epimorphism iff $f(X)$ were a dense subset of (Y, H).

(v) In the categories listed in (ii), a regular monomorphism is a homeomorphism of the first space on a subspace of the second, whereas a regular epimorphism is a homeomorphism composed with a quotient mapping of the first space on a quotient.

(vi) The spaces of (ii) are well-powered and cowell-powered. Each has an initial object (based on a null set) and a final object (based on a one-element set) but no zero.

(vii) *PS* and *PTOP* both have a zero object. They are pointed categories.

CHAPTER FOUR

TYPES OF FUNCTORS

4.1 FULL, FAITHFUL, DENSE, EMBEDDING FUNCTORS

Just as morphisms in a category connect objects, we now consider the functors that connect different categories. We have already encountered one, the underlying set functor U for a concrete category. The following definition generalizes that and other cases:

Definition 4.1.1 Given categories C and D, we say that F is a *(covariant) functor* from C to D, and indicate this by the notation $F: C \to D$, if the following are true: (i) for each A in $O(C)$, $F(\mathrm{A})$ is a uniquely determined element of $O(D)$; (ii) for A, B from $O(C)$ and any f of $\hom_C(\mathrm{A}, \mathrm{B})$, $F(f)$ is a uniquely determined element of $\hom_D(F(\mathrm{A}), F(\mathrm{B}))$; (iii) when f and g of $M(C)$ are such that $\mathrm{dom}(f) = \mathrm{cod}(g)$ [so that $f \cdot g$ is defined in $M(C)$], $F(f \cdot g) = F(f) \cdot F(g)$; (iv) $F(1_{\mathrm{A}}) = 1_{F(\mathrm{A})}$ for each A of $O(C)$.

We call F a *contravariant functor* from C to D if F is a covariant functor from C to D^{op}; this is equivalent to changing, in the preceding set of conditions, (ii) and (iii) to (ii′) and (iii′) where (ii′) says, for A, B, from $O(C)$, and any f in $\hom_C(\mathrm{A}, \mathrm{B})$, $F(f)$ is a uniquely determined element of $\hom_D(F(\mathrm{B}), F(\mathrm{A}))$; and (iii′) says, when f, g are from $M(C)$ with $\mathrm{dom}(f) = \mathrm{cod}(g)$, $F(f \cdot g) = F(g) \cdot F(f)$.

Hereafter, whenever we talk of a functor without a qualifying prefix we understand it to mean a covariant functor only.

Definition 4.1.2 A functor $F: C \to D$ is called (i) *faithful*/(ii) *full* if [for any A, B from $O(C)$] F maps $\hom_C(\mathrm{A}, \mathrm{B})$ in $\hom_D(F(\mathrm{A}), F(\mathrm{B}))$ (i) injectively/

(ii) surjectively. F is called *dense* if each D of $O(D)$ is isomorphic with some $F(\mathrm{A})$ for an A of $O(C)$; and F is called an *embedding* if F is an injective mapping of the class $M(C)$ in $M(D)$.

For an embedding F, F has to be faithful and one-to-one on objects also.

If F: $C \to D$, G: $D \to K$, and H: $K \to L$ are all functors, there is a natural way we can compose these to get functors $G \cdot F$: $C \to K$, $H \cdot G$: $D \to L$, and $H \cdot (G \cdot F)$ and $(H \cdot G) \cdot F$ both from C to L. It is easily seen that the triple products $(H \cdot G) \cdot F$ and $H \cdot (G \cdot F)$ are the same, and that $G \cdot F$ is covariant when both F and G are either covariant or both contravariant, whereas if one is covariant and the other contravariant, the composition $G \cdot F$ is a contravariant functor. We note some typical examples of these functors.

Example 1. The underlying set functor for a concrete category: For a concrete category $C = (U_C, \mathrm{hom}_C, \cdot_C)$, the U_C: $C \to S$ is a faithful functor. In fact, the existence of such a faithful functor is characteristic for a concrete category.

Example 2. Inclusion functors: When C is a subcategory of D, the inclusion functor I: $C \to D$, which takes objects or morphisms in C to themselves considered as objects or morphisms of D, is an embedding functor. This functor is also full if the subcategory is a full subcategory.

Example 3. Quotient functors: When E is a congruence on a category C, we defined (in Section 2.2) a quotient category C/E; there is a "quotient functor" ϕ: $C \to C/E$, with $\phi(\mathrm{A}) = \mathrm{A}$ for each object A of C and $\phi(f)$ equal to the E-class f^E containing f for each morphism f of C; this is a full functor, which we refer to as the *canonical quotient functor*.

Example 4. The hom functors: For any category C we form first the product $C^{\mathrm{op}} \times C$; from this product category to the category S of sets there is a functor called the *hom functor*; hom: $C^{\mathrm{op}} \times C$ is defined as follows: for an object (A, B) from $C^{\mathrm{op}} \times C$, $\mathrm{hom}(\mathrm{A}, \mathrm{B}) = \mathrm{hom}_C(\mathrm{A}, \mathrm{B})$ [in $O(S)$]. For a morphism (f, g) from (A, B) to (A′, B′) in $C^{\mathrm{op}} \times C$ [that is, for f and g from $\mathrm{hom}_C(\mathrm{A}', \mathrm{A})$ and $\mathrm{hom}_C(\mathrm{B}, \mathrm{B}')$], $\mathrm{hom}(f, g) = g(\ )f$: $\mathrm{hom}_C(\mathrm{A}, \mathrm{B}) \to \mathrm{hom}_C(\mathrm{A}', \mathrm{B}')$ is determined by the rule $[g(\ )f](\theta) = g \cdot \theta \cdot f$. It is not difficult to verify that this indeed gives a functor from $C^{\mathrm{op}} \times C$ to S. Further, for any chosen object A from C we have a covariant functor $(\mathrm{A}, -)$: $C \to C^{\mathrm{op}} \times C$ and a contravariant functor $(-, \mathrm{A})$: $C \to C^{\mathrm{op}} \times C$ defined by $(\mathrm{A}, -)(\mathrm{B}) = (\mathrm{A}, \mathrm{B})$ and $(\mathrm{A}, -)(g) = (1_{\mathrm{A}}, g)$ for any B in $O(C)$ and g in $\mathrm{hom}_C(\mathrm{B}, \mathrm{B}')$; while for the same B and g, $(-, \mathrm{A})(\mathrm{B}) = (\mathrm{B}, \mathrm{A})$ and $(-, \mathrm{A})(g) = (g^{\mathrm{op}}, 1_{\mathrm{A}})$ where g^{op} is in $\mathrm{hom}_{C^{\mathrm{op}}}(\mathrm{B}', \mathrm{B})$ and corresponds to g in $\hom_C(B, B')$. By combining one of these with the earlier hom functor we get two special hom functors from C to S, a covariant one equal to hom(A, $-$) and a contravariant one equal to hom ($-$, A), determined by A of $O(C)$, both being functors from C to S. We call these the *covariant/contravariant representation functors determined by* A.

4.2 REFLECTION AND PRESERVATION OF CATEGORICAL PROPERTIES

When we look at the underlying set functor U for a category of algebras, say FDA, we know that a morphism f in FDA is a monomorphism or an epimorphism if and only if $U(f)$ is also such in S [that is, $U(f)$ is injective or surjective]. These forms of results are what are being expressed here in general terms.

We call $P(A, \ldots; f, \ldots)$ a categorical property when $A, \ldots$ and $f, \ldots$ are symbols standing for objects and morphisms and P is some statement involving such objects and morphisms that makes sense (that is, would be true or false) in any category—for example, (A is an initial object), or (f is an epimorphism), or (e is an equalizer of the set (f_j: j in J)).

Definition 4.2.1 Given a functor F: $C \to D$, we say that F *preserves the property* P(A, ...; f, ...) if, given any set of objects (A, ...) and morphisms (f, ...) from C as possible values in the expression for P(A, ...; f, ...), [P(A, ...; f, ...) is true in C] always implies that [$P(F(A), \ldots; F(f), \ldots)$ is true in D]; and we say that F *reflects the property* P if, given the objects (A, ...) and morphisms (f, ...) from C, [$P(F(A), \ldots; F(f), \ldots)$ is true in D] implies that [P(A, ...; f, ...) is true in C].

With these conventions we can now look at a number of such cases.

Theorem 4.2.1 *(i) Any functor F: $C \to D$ preserves identities, sections, retractions, isomorphisms, and commutative triangles.*

(ii) A faithful functor F: $C \to D$ reflects monomorphisms, epimorphisms, bimorphisms, constant morphisms, coconstant morphisms, zero morphisms, and commutative triangles.

(iii) A full and faithful functor F: $C \to D$ reflects sections, retractions, and isomorphisms.

(iv) A full, faithful, and dense functor F: $C \to D$ preserves monomorphisms, epimorphisms, bimorphisms, constant morphisms, coconstant morphisms, and zero morphisms.

PROOF. We start with observing that duality would prove, for each part proved, a dual part; so that when, for instance, monomorphisms and epimorphisms are involved, the proof for one suffices.

(i) The definition of a functor requires that $F(1_A) = 1_{F(A)}$ for any A in

$O(C)$, while for a given f in a $\hom_C(A, B)$, the existence of a g in $\hom_C(B, A)$ with $g \cdot f = 1_A$ would mean that there is an $F(g)$ in $\hom_D(F(B), F(A))$ for $F(f)$ in $\hom_D(F(A), F(B))$, with $F(g) \cdot F(f) = F(g \cdot f) = 1_{F(A)}$. Thus F preserves sections, hence by duality, retractions, and similarly isomorphisms. When f, g and h are morphisms forming the side of a commutative triangle in C with $g \cdot f = h$, clearly $F(g) \cdot F(f) = F(h)$ and $F(f)$, $F(g)$, $F(h)$ form the sides of an image commutative triangle in D.

(ii) Given f in $\hom_C(A, B)$, and morphisms g, h from $\hom_C(C, A)$ for some C of $O(C)$, if $F(f)$ is a monomorphism, $f \cdot g = f \cdot h$ would imply $F(f) \cdot F(g) = F(f) \cdot F(h)$ for $F(g)$, $F(h)$ from $\hom_D(F(C), F(A))$. Hence, $F(g) = F(h)$; but F being faithful, this implies that $g = h$ also. Thus we have proved that f is a monomorphism in C. Dually we can prove that F reflects epimorphisms, hence also bimorphisms.

If $F(f)$ were a constant morphism, with the choice of g, h as above, $F(f) \cdot F(g) = F(f) \cdot F(h)$ would follow from assuming that $F(f)$ is a constant morphism, or $F(f \cdot g) = F(f \cdot h)$; again, using the faithfulness of F, we get $f \cdot g = f \cdot h$. Thus f is a constant morphism. Dually we prove that F reflects coconstant morphisms, hence zero morphisms. Finally, if f, g, and h are morphisms that form the sides of a triangle (in a diagram) in C, with $F(g) \cdot F(f) = F(h)$ in D, faithfulness of F gives $g \cdot f = h$ in C; thus F reflects commutative triangles.

(iii) If we assume now that F is full and faithful, f is in $\hom_C(A, B)$, and $F(f)$ is a section in D, there must be a g' in $\hom_D(F(B), F(A))$ with $g' \cdot F(f) = 1_{F(A)} = F(1_A)$; since F is full, there is a g in $\hom_C(B, A)$ for which $F(g) = g'$; hence $F(g \cdot f) = F(1_A)$, from which follows $g \cdot f = 1_A$, since F is faithful. Thus f is also a section. Similarly we prove that F reflects retractions and isomorphisms.

(iv) To prove this last result, assuming that $F: C \to D$ is full, faithful, and dense, let f in $\hom_C(A, B)$ and g', h' be any morphisms in D from $\hom_D(D, F(A))$ for some object D of D. Since F is dense, we can find a C in $O(C)$ and an isomorphism $j: F(C) \to D$ in D. Then, for the morphisms $g' \cdot j$, $h' \cdot j$ from $\hom_D(F(C), F(A))$ we can find g, h from $\hom_C(C, A)$ with $F(g) = g' \cdot j$ and $F(h) = h' \cdot j$.

Let us first assume that f is a monomorphism in C; to prove that $F(f)$ is a monomorphism in D, we have to show that $F(f) \cdot g' = F(f) \cdot h'$ would imply that $g' = h'$; but $[F(f) \cdot g' = F(f) \cdot h'] \Rightarrow [F(f) \cdot g' \cdot j = F(f) \cdot h' \cdot j] \Rightarrow [F(f) \cdot F(g) = F(f) \cdot F(h)] \Rightarrow [F(f \cdot g) = F(f \cdot h)] \Rightarrow$ [(since F is faithful), $f \cdot g = f \cdot h] \Rightarrow [g = h$, since f is a monomorphism$] \Rightarrow [g' \cdot j = F(g) = F(h) = h' \cdot j] \Rightarrow [g' \cdot j \cdot j^r = h' \cdot j \cdot j^r$, since j has an inverse $j^r] \Rightarrow [g' = h']$. Thus we have shown that F preserves monomorphisms; dually it preserves epimorphisms, and hence also bimorphisms.

Assume next that f is a constant morphism in C; to prove that $F(f)$ is a constant morphism in D it suffices to show that $F(f) \cdot g' = F(f) \cdot h'$, since g', h' are arbitrary morphisms from $\hom_D(\mathrm{D}, F(\mathrm{A}))$. Since f is a constant morphism, we have $f \cdot g = f \cdot h$; hence $F(f \cdot g) = F(f \cdot h)$ or $F(f) \cdot g' \cdot j = F(f) \cdot h' \cdot j$; right-multiplying these by j^r, we get $F(f) \cdot g' = F(f) \cdot h'$. Hence we have proved that F preserves constant morphisms; dually it preserves coconstant, and hence zero, morphisms. □

That a full, faithful, dense functor may not reflect identities is seen from the example: let C be the category whose objects are finite ordered sets [of the form $(a_1, \ldots, a_n)$ for some integer $n > 0$] and morphisms are the isomorphisms of the form (a, b): $(a_1, \ldots, a_n) \to (b_1, \ldots, b_n)$, where $(a, b)(a_i) = b_i$ for each i. Let D have the set N of integers greater than 0 for the class of objects and identities on these as morphisms. The obvious functor F: $C \to D$ is faithful, full, and dense; but there are isomorphisms in C other than identities that go by F to identities in D.

The covariant functor $\hom(\mathrm{A}, -)$: $C \to S$ for an object A of C is faithful iff A is a separator in C, and this functor always preserves monomorphisms. We call A a *projective object* in C if this functor preserves epimorphisms; in other words, given any epimorphism f from $\hom_C(\mathrm{B}, \mathrm{C})$, and an h from $\hom_C(\mathrm{A}, \mathrm{C})$, there is a g in $\hom_C(\mathrm{A}, \mathrm{B})$ such that $h = f \cdot g$.

Dually, A is called an *injective object* in C if, given a monomorphism f from $\hom_C(\mathrm{B}, \mathrm{C})$, and an h of $\hom_C(\mathrm{B}, \mathrm{A})$, there is a g of $\hom_C(\mathrm{C}, \mathrm{A})$ with $h = g \cdot f$; in other words, the covariant functor $\hom(-, \mathrm{A})$ from C^{op} to S preserves epimorphisms. In the category of modules over a ring, the projective/injective objects are the usual projective/injective modules.

4.3 THE FEEBLE FUNCTOR AND REVERSE QUOTIENT FUNCTOR

The (covariant) functor as we defined it, F: $C \to D$, is essentially a mapping relation of the class $M(C)$ in $M(D)$ that preserves the partial binary operation of composition in these classes. If we replace this by more general relations we get variations of the notion. The one we consider now replaces the mapping relation by a cofull relation with a suitable "preservation of composition" condition.

Definition 4.3.1 Given categories C and D, we say that F is a *feeble functor* from C to D, in symbols F: $C \dashrightarrow D$, if the following are true:

(i) for each A in $O(C)$, $F(\mathrm{A})$ is a uniquely determined element of $O(D)$;

(ii′) for any A, B from $O(C)$ and any f from $\hom_C(\mathrm{A}, \mathrm{B})$, $F(f)$ is a nonnull subset of $\hom_D(F(\mathrm{A}), F(\mathrm{B}))$;

(iii′) when f, g of $M(C)$ are such that $\mathrm{dom}(f) = \mathrm{cod}(g)$ (so that $f \cdot g$ is defined in $M(C)$), $F(f \cdot g)$ contains $F(f) \cdot F(g)$ [$=(f' \cdot g'$: f' in $F(f)$, g' in $F(g))$];

(iv′) $F(1_A)$ contains $1_{F(A)}$, for each A of $O(C)$;
(v) $[F(f) \cap F(k) \neq \emptyset] \Rightarrow [F(f) = F(k)]$.

Dually (reversing arrows in D) we can define a *contravariant feeble functor*. The one we have defined is called *covariant*.

A very natural and in a sense typical feeble functor arises from the formation of quotient categories. If C is a category with E a congruence on it, and ϕ denotes the canonical quotient functor from C to C/E, we can define a feeble functor χ from C/E to C called the reverse of ϕ (and a *reverse quotient functor*) by setting $\chi(A) = A$ for each A from $O(C/E) = O(C)$ and $\chi(f^E) = [f' : f' \text{ in } f^E]$ for any E-class of morphisms considered as a morphism in C/E. It is easy to check that this gives a feeble functor.

Definition 4.3.2 A feeble functor F: C- - - → D is called (i) *faithful* if $F(f) = F(g)$ implies $f = g$ for any f, g from $M(C)$; (ii) *full* if, for any A, B from $O(C)$ and any h from $\hom_D(F(A), F(B))$, there is some f in $\hom_C(A, B)$ such that h belongs to $F(f)$; (iii) *dense* if each B of $O(D)$ is isomorphic to some $F(A)$, A from $O(C)$; and (iv) a *feeble embedding* if it is one-to-one on objects and a faithful feeble functor.

Given feeble functors F: C --→ D and G: D --→ K, we can define a composite feeble functor $G \cdot F$: C --→ K by setting $(G \cdot F)(A) = G(F(A))$ for each A from $O(C)$ and $(G \cdot F)(f) = U[G(g): g \text{ in } F(f)]$ for each f from $M(C)$. This composition operation is associative when the triple products are defined, $H \cdot (G \cdot F) = (H \cdot G) \cdot F$.

For feeble functors between the same pair of categories C and D we can also define a relation of order $\leq$ by setting $[F \leq F']$ iff [for each f in $M(C)$: $F(f) \subseteq F'(f)$]. The functors themselves can obviously be treated as special feeble functors in which the image sets $F(f)$ are all one-element sets. This order relation applied to functors between C and D clearly becomes the relation of equality.

To relate the feeble functors with functors, we next describe two relations on $M(D)$ determined by a feeble functor F: C --→ D: the first $R(F) = 1_{M(D)} \cup F \cdot F^r = [(g, g')$: either $g = g'$ or, for some f of $M(C)$, g and g' belong to $F(f)]$. The second is the congruence on D generated by $R(F)$; thus this $E(F)$ equals the transitive closure of $S(F)$, a reflexive and symmetric relation on $M(D)$ given by $S(F) = [(h, h')$: there is a finite set of pairs $(g_1, g_1'), \ldots, (g_r, g_r')$ all in $R(F)$ such that $h = g_1 \cdot g_2 \cdot \cdots \cdot g_r$ and $h' = g_1' \cdot g_2' \cdot \cdots g_r']$. If ϕ is the canonical quotient functor from D on $D/E(F)$, it is clear that $F\# = \phi \cdot F$ is a functor from C to $D/E(F)$; we call this $F\#$ the functor associated with the feeble functor F.

Definition 4.3.3 A feeble functor F that can be factored in the form $F = \chi \cdot F^*$, where F^* is a functor and χ is a reverse quotient functor (the reverse of a ϕ), is called a *closed feeble functor*. In particular, any reverse quotient functor is a closed feeble functor.

Lemma 4.3.1 *(a) Given feeble functors $F: C \dashrightarrow D$, $G: D \dashrightarrow K$, the composite feeble functor $G \cdot F: C \dashrightarrow K$ is faithful, full, or an embedding feeble functor when each of F and G is such a functor. A quotient functor, as well as a reverse quotient (feeble) functor, is full and dense.*

(b) A feeble functor $F: C \dashrightarrow D$ is closed iff there is a congruence E on D containing E(F) such that [an E-class E(g) of M(D) meets F(f) for an f in M(C)] implies [E(g) = F(f)]; in that case $F = \chi \cdot F\#$, with $F\# = \phi \cdot F$ a functor; there is a least closed feeble functor F^* containing any feeble functor $F: C \dashrightarrow D$ (its closure), namely $F^* = \chi \cdot \phi \cdot F$, if $\phi: D \rightarrow D/E(F)$ and χ is its reverse.

(c) Given a feeble functor $F: C \dashrightarrow D$ and the associated functor $F\#: C \dashrightarrow D/E(F)$ ($F\# = \phi \cdot F$), we have the following implications: (i) [F is full] $\Rightarrow$ [F is closed]; (ii) [F is full] $\Rightarrow$ [F# is full]; (iii) [F# is full and F is closed] $\Rightarrow$ [F is full]; (iv) [F# is faithful] $\Rightarrow$ [F is faithful]; (v) [F is closed and faithful] $\Rightarrow$ [F# is faithful]; (vi) [F is dense] $\Rightarrow$ [F# is dense].

PROOF. (a) If F and G are both faithful, let f, f' be from $M(C)$ with $f \neq f'$; then $F(f)$ and $F(f')$ are disjoint. Hence $g \neq g'$ if g, g' are from $F(f)$ and $F(f')$, respectively; hence $G(g)$ and $G(g')$ would be disjoint for these. Finally then $\cup[G(g)\colon g \text{ in } F(f)]$ and $\cup[G(g')\colon g' \text{ in } F(f')]$ are also disjoint; that is, $G \cdot F(f)$ and $G \cdot F(f')$ are disjoint. This implies that $G \cdot F$ is faithful.

Next, if F and G are both full, for objects A, B of C and any morphism h from $\hom_K(G \cdot F(A), G \cdot F(B))$ there is a g in $\hom_D(F(A), F(B))$ with $G(g)$ containing h; and for this g there is an f in $\hom_C(A, B)$ with $F(f)$ containing g; thus finally $G \cdot F(f)$ contains h, or $G \cdot F$ is also full.

When F and G are both embedding functors, they are faithful and one–one on objects; then by our earlier result, $G \cdot F$ is also faithful, and clearly also one–one on objects. A quotient functor or its reverse is surely full and dense.

(b) Let $F: C \dashrightarrow D$ be a closed feeble functor, so $F = \chi \cdot F\#$, where $F\#$ is a functor from C to D/E, E is a congruence on D and χ is the reverse of the canonical quotient functor $\phi: D \rightarrow D/E$. In this case $\phi \cdot \chi = 1_{D/E}$, so that $\phi \cdot F = \phi \cdot \chi \cdot F\# = F\# =$ a functor. If $g' \in E(g)$, $g \in F(f)$ for an f of $M(C)$ and for g, g' from $M(D)$, $\phi(g') = \phi(g)$ (since $\phi: D \rightarrow D/E$ is the canonical quotient functor) and $\phi(g)$ is in $\phi \cdot F(f) = F\#(f)$. Hence it follows that $\phi(g')$ is in $F\#(f)$ and g' is in $\chi \cdot \phi(g') \subseteq \chi \cdot F\#(f) = F(f)$; that is, when $E(g)$ and $F(f)$ meet, $E(g)$ is contained in $F(f)$; but surely $F(f)$ is also contained in $E(g)$ when g is in $F(f)$, since any g'' of $F(f)$ satisfies $\phi(g'') = F\#(f) = \phi(g)$ [seeing that $F\#(f)$ is a one-element set, $F\#$ being a functor]. Thus $E(g) = F(f)$ when $E(g)$ meets an $F(f)$. Conversely, let us now assume that E is a congruence on D containing $E(F)$, where $F: C \dashrightarrow D$ is a given feeble functor, and that any E-class of $M(D)$ that meets an $F(f)$ equals it; if $\phi: D \to D/E$ is the canonical quotient functor and χ is its reverse, we first show that $F\# = \phi \cdot F$ is a functor from C to D/E; for if f is any morphism from C, g, g' are in $F(f)$ implies (g, g') belongs to $R(F)$ and so also to $E(F)$ and finally to E, since E contains $E(F)$. Hence $\phi(g) = \phi(g')$, or $F\#(f)$ is a one-element set of $M(D/E)$. This implies that the feeble functor $\phi \cdot F$ is a functor. Next we show that $F = \chi \cdot F\#$. For any f from $M(C)$, if g is in $F(f)$, the E-class $E(g)$ containing g meets $F(f)$, so must be equal to it; thus $E(g) = F(f)$. But $\chi \cdot F\#(f) = \chi \cdot \phi \cdot F(f) = \chi \cdot \phi \cdot E(g) = E(g)$ because E is the congruence defining the canonical functor ϕ. Thus $F = \chi \cdot F\#$.

Next, given a feeble functor $F: C \dashrightarrow D$, if $\phi: D \to D/E(F)$ is the canonical quotient functor and χ is its inverse, we set $F^* = \chi \cdot \phi \cdot F = \chi \cdot F\#$, with $F\# = \phi \cdot F$; if we show that $F\#$ is a functor, then it would follow that F is a closed feeble functor, since it factors as $F = \chi \cdot F\#$ for χ a reverse quotient functor. Given any f in $M(C)$ and any elements g, g' from $F(f)$, we know that (g, g') belongs to $R(F)$ and so also to $E(F)$, which contains $R(F)$. Hence $\phi(g) = \phi(g')$; thus $\phi \cdot F(g)$ is a one-element set of $M(D/E)$ for each f of $M(C)$. This means that $F\#$ is a functor. Evidently F^* contains F; if a closed feeble functor $H: C \dashrightarrow D$ contains F, $H = \chi' \cdot \phi' \cdot H$, where ϕ' is a canonical quotient functor $\phi': D \to D/E'$ for a congruence E' on $M(D)$, χ' is its reverse, and $H\# = \phi' \cdot H$ is a functor. If (g, g') is in $R(F)$, (g, g') is in E', too. Since this is obvious if $g = g'$, assume that g, g' are from an $F(f)$, f in $M(C)$; then, since H contains F, $H(f)$ contains $F(f)$ and so g, g' are both in $H(f)$; then $\phi'(g) = \phi'(g')$ must be in $\phi' \cdot H(f) = H\#(f)$, which is a one-element set of $M(D/E')$, since $H\#$ is a functor. Thus (g, g') must be in E'; since E' is a congruence on D containing $R(F)$, it would also contain $E(F)$, which is the smallest congruence containing $R(F)$.

(c) (i) If F is full, when the $E(F)$-class containing a g of $M(D)$ meets an $F(f)$, for f from $M(C)$, it surely contains $F(f)$ [from the definition of $R(F)$ and $E(F)$], while if g' is in this $E(F)$-class, $(g,g') \in E(F)$ means that there is a sequence $g = g_1, \ldots, g_{r+1} = g'$ from g to g' of elements of $M(D)$ such that the pairs (g_j,g_{j+1}), for $j = 1, \ldots r$, are all in $S(F)$. From (g_1,g_2) in $S(F)$ follows that $g_1 = h_1 \cdot \cdots \cdot h_k$ and $g_2 = h_1' \cdot \cdots \cdot h_k'$ where, for $i = 1, \ldots, k$, either $(h_i = h_i')$ or h_i and h_i' both belong to some $F(f_i)$ for f_i from $M(C)$; but, since F is full, even the first possibility $(h_i = h_i')$ implies that h_i and h_i' are both in some $F(f_i)$. Hence there is an $f^* = f_1 \cdot \cdots \cdot f_k$ of $M(C)$ such that $F(f^*)$ contains both g_1 and g_2. Similarly some $F(f')$ contains both g_2 and g_3; but then as $F(f^*)$ and $F(f')$ have a common element (g_2), they must be equal. Thus $F(f^*)$ contains g_1, g_2, g_3, and so on up to $g_{r+1} = g'$. Since the $E(F)$-class does contain a g' from $F(f)$, that such a g' is in $F(f)$

and in $F(f^*)$ would imply $F(f) = F(f^*)$. Hence every g' of the $E(F)$-class belongs to $F(f)$. These prove that F is a closed feeble functor.

(ii) If F is full, since ϕ is also full, $F\# = \phi \cdot F$ is full, by (a).

(iii) If F is closed and $F\#$ is full, $F = \chi \cdot F\#$ is full, because χ is full.

(iv) If $F\#$ is faithful, for any f, f' from $M(C)$, $F(f) = F(f')$ implies $\phi \cdot F(f) = \phi \cdot F(f')$ or $F\#(f) = F\#(f')$, which in turn implies that $f = f'$ (since $F\#$ is faithful). Hence F must also be faithful, from the definition of faithfulness for feeble functors.

(v) If we assume that F is closed and faithful, and if $F\#(f) = F\#(f')$ for some f, f' from $M(C)$, then $\chi \cdot F\#(f) = \chi \cdot F\#(f')$ or $F(f) = F(f')$, which implies that $f = f'$. Hence $F\#$ is seen to be faithful.

(vi) If F is dense, since ϕ is one–one on objects, $F\# = \phi \cdot F$ is seen to be dense too. □

EXERCISES AND REMARKS ON CHAPTER 4

Example 1. Check that the inclusion functor I: $F^*D^*A \to FDA$ from the category of F^*D^*-algebras and their F^*-homomorphisms to the category of FD-algebras and their F-homomorphisms, when $F^* \supseteq F$ and $D^* \supseteq D$, is a faithful functor that need not be full, dense, or an embedding functor.

Example 2. Check that the "free algebra" functors $S \to FA$, $S \to FDA$ suggested by Lemmas 1.3.2 and 1.3.3 are embedding functors, but neither dense nor full.

Example 3. The categories formed by neighborhood spaces, convergence spaces, closure spaces, topological spaces, each with continuous maps as the morphisms, have functors connecting them: $NS \to CS \to ClS \to T \to NS$ suggested by Lemmas 1.4.1 and 1.4.2. Check that these functors are faithful, full, and dense.

Example 4. Check that the inclusion functors between categories of T_4-, T_3-, T_2'-, T_1'-, T_0-, CR-spaces, with continuous maps for morphisms, as suggested in Lemma 1.4.5, are full embedding functors.

Example 5. The inclusion functors $SM \to SU$, $QM \to QU$, $M \to U$ are faithful, but neither dense, full, nor embedding functors.

Example 6. In the category of semiuniform spaces with uniform maps as morphisms we have a subcategory whose objects are complete semiuniform spaces. We designate these complete ones by adding a C at the beginning; thus we have CSU, CQU, CU, CUT_0, etc., all of which are supposed to be full subcategories of SU. Theorem 1.6.1 then gives us the functors $SU \to CSUT_0$ and $QSU \to CUT_0$. Denoting the first by C^*, we can treat the second as the same functor restricted to the subcategory QSU of SU, therefore denoted by the same symbol. These functors are dense (any complete T_0 semiuniform space is "isomorphic" with its own complete associate). They are not full nor (generally) faithful, as can be seen by looking at the metric example of completing the rationals by the reals, and by starting with a non-T_0 semiuniform space for the completion process. The functor would be faithful if the domain were restricted to T_0 uniform spaces (which are Hausdorff).

Example 7. There is a functor P: $SU \rightarrow PO$ from semiuniform spaces to preordered sets, the preorder on X being defined from a semiuniformity (U, J) on X by taking the intersection of the relations $[U(j)^r: j \text{ in } J]$. We shall discuss this example in more detail later. This functor is full and dense, but neither faithful nor embedding.

Example 8. Examples of identifying functors: There is one Q_1: $PO \rightarrow O$ from the category of preordered sets (and monotone maps) to the subcategory of ordered sets, $Q_1((X, \leq)) = (X/E, \leq/E)$ where E is the equivalence $[\leq \cap \geq]$ on X; and $Q_1(g)$, for a monotone map g from $(X, \leq)$ to $(Y, \leq)$, is g/E: $(X/E, \leq/E) \rightarrow (Y/E', \leq/E')$ where $g/E(x^E) = (g(x))^{E'}$. There is a similar one, Q_2: $T \rightarrow T_0$, obtained by setting $Q_2(X, \mathbf{G}) = (X/E, \mathbf{G}/E)$ where E is the equivalence on X defined by the topology $\mathbf{G}$ as xEx' iff every open set containing one of x or x' always contains the other—what is usually called *topological equivalence* (or *T-equivalence*). $\mathbf{G}/E = [H/E : H \text{ in } \mathbf{G}]$.This quotient space is a T_0-space, called the T_0-associate of (X, G). Each of Q_1 and Q_2 is full and dense, but not a faithful nor an embedding functor.

CHAPTER FIVE

NATURAL TRANSFORMATIONS AND EQUIVALENCES

5.1 NATURAL TRANSFORMATIONS AND THEIR COMPOSITIONS

Even as morphisms relate objects and functors relate categories, we have now the natural transformations that relate functors. We start with a classic example. When V denotes the category of vector spaces over the reals, we have a contravariant functor $C: V \to V$ given by $C(X, +) = (X^*, +)$, where $X^* =$ [the set of linear transformations $x^*: (X, +) \to (R, +)$] with the vector operations for these defined naturally to make it a vector space $(X^*, +)$, and $C[h: (X, +) \to (Y, +)] = h^*: (Y^*, +) \to (X^*, +)$ is such that $x^* = h^*(y^*)$ for a y^* of Y^* is specified by $x^*(x) = y^*(h(x))$, for any x of X. From C we then get a covariant functor C^2 (or $C \cdot C$) of V in V. For each $(X, +)$ in $O(V)$, there is a morphism $\eta(X, +): (X, +) \to C^2(X, +)$ given by $[\eta(X, +)](x) = x^{**}$ in $C^2(X, +)$, where this element x^{**} is specified by setting $x^{**}(x^*)$, for any x^* of $C(X, +)$, to be the same as $x^*(x)$. Further, this association of a morphism in V to each object of V satisfies the condition that if $f: (X, +) \to (Y, +)$ is a morphism of V, $\eta(Y, +) \cdot f = C^2(f) \cdot \eta(X, +)$. These properties would translate, under the definition that follows, to the statement that η is a natural transformation from the functor 1_V to C^2.

Definition 5.1.1 Given a pair of functors F_1, F_2, both from a category A to a category B, we say that η is a *natural transformation* from F_1 to F_2, and write for this $\eta: F_1 \to F_2$ if, for each object A from A, $\eta(\mathrm{A})$ is a uniquely associated morphism in B from $F_1(\mathrm{A})$ to $F_2(\mathrm{A})$ such that, when $f: \mathrm{A} \to \mathrm{A}'$ is any morphism in A, we have the commutativity condition $F_2(f) \cdot \eta(A) = \eta(A') \cdot F_1(f)$. (See Figure 5.1.) The natural transformation is called a *natural equivalence* if, for each object A of A, $\eta(A)$ is an isomorphism in B.

We can check that the η defined earlier is a natural transformation from 1_V to C^2. If we restrict V to consist of only finite-dimensional vector spaces, it is also known that this would be a natural equivalence.

These natural transformations can be composed in two different ways. For the first, *horizontal composition*, we consider three functors $F_i (i = 1, 2, 3)$, all from A to B, and two natural transformations $\eta: F_1 \to F_2$ and $\lambda: F_2 \to F_3$. The composite $\lambda \circ \eta: F_1 \to F_3$ is then determined by the rule; For any object A of A, $(\lambda \circ \eta)(A) = \lambda(A) \cdot \eta(A)$ [a composition in $M(B)$ of morphisms].

This composition is associative. For, given one more natural transformation $\mu: F_3 \to F_4$ (and one more functor $F_4: A \to B$), the two triple products $(\mu \circ \lambda) \circ \eta$ and $\mu \circ (\lambda \circ \eta)$ are equal. This equality follows directly from the associative law for morphisms in B. That $\lambda \circ \eta$ is a natural transformation from F_1 to F_2 is seen from Figure 5.1. For a typical f in $M(A)$, $[(\lambda \circ \eta)(A')] \cdot F_1(f) = \lambda(A') \cdot [\eta(A') \cdot F_1(f)] = [\lambda(A') \cdot F_2(f)] \cdot \eta(A) = F_3(f) \cdot [\lambda(A) \cdot \eta(A)] = F_3(f) \cdot (\lambda \circ \eta)(A)$.

Figure 5.1

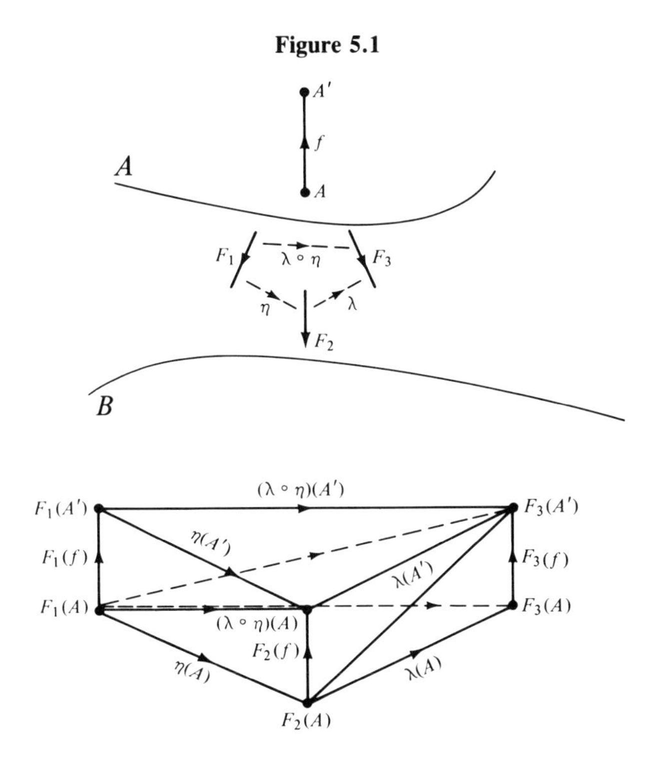

An obvious example of a natural equivalence is the "identity" 1_F from $F: A \to B$ to F itself, $(1_F)(A)$ being $1_{F(A)}$ for each A from $O(A)$.

The second, *vertical composition* of natural transformations, links together a natural transformation $\eta: F_1 \to F_2$ with another $\delta: G_1 \to G_2$, where F_1, F_2 are both functors from A to B, and G_1, G_2 are both functors from B to C to give the natural transformation $(\delta \# \eta): (G_1 \cdot F_1) \to (G_2 \cdot F_2)$. For an A from $O(A)$, $\eta(A): F_1(A) \to F_2(A)$ is a morphism in B, and since δ is a natural transformation from G_1 to G_2, there is a commutative relation $\delta F_2(A) \cdot G_1\eta(A) = G_2\eta(A) \cdot \delta F_1(A)$ in C. This common morphism from $G_1F_1(A)$ to $G_2F_2(A)$ we denote by $(\delta \# \eta) \cdot (A)$. Similarly, we define $(\delta \# \eta)(A')$ for A′ from $O(A)$. To see that this gives indeed a natural transformation, we have to check the commutativity condition: $(\delta \# \eta)(A') \cdot G_1F_1(f) = G_2F_2(f) \cdot (\delta \# \eta)(A)$. Using the fact that η is also a natural transformation, we get the sequence of equalities $[(\delta \# \eta)(A')] \cdot G_1F_1(f) = \delta F_2(A') \cdot [G_1\eta\ (A') \cdot G_1F_1(f)] = [\delta F_2(A') \cdot G_1F_2(f)] \cdot G_1\eta(A) = G_2F_2(f) \cdot [\delta F_2(A) \cdot G_1\eta(A)] = G_2F_2(f) \cdot (\delta \# \eta)(A)$. (See Figure 5.2.)

This vertical composition is also an associative operation. If, in addition to the η and δ of the last paragraph, we had functors H_1, H_2 from C to D and a natural transformation $\theta: H_1 \to H_2$, the associative law for the triple product would follow if we check that for any object A from A, the two forms of the triple product associate the same morphism in D; this we check: $[(\theta \# \delta) \# \eta](A) = (\theta \# \delta)(F_2(A)) \cdot H_1G_1\eta(A) = \theta G_2F_2(A) \cdot H_1\delta F_2(A) \cdot H_1G_1\eta(A) = \theta G_2F_2(A) \cdot H_1(\delta \# \eta)(A) = [\theta \# (\delta \# \eta)](A)$. (See Figure 5.3.)

The two types of composition are connected by an "interchange law." If there are three functors F_i between A and B and three functors G_i between B and C ($i = 1, 2, 3$), and natural transformations $\eta: F_1 \to F_2$, $\eta': F_2 \to F_3$, $\delta: G_1 \to G_2$, and $\delta': G_2 \to G_3$, this law asserts that the two natural transformations $(\delta' \# \eta') \circ (\delta \# \eta)$ and $(\delta' \circ \delta) \# (\eta' \circ \eta)$ from (G_1F_1) to (G_3F_3) are identical. We prove this by checking that these give the same value [a morphism in C from $G_1F_1(A)$ to $G_3F_3(A)$] for each object A of A: for $(\delta' \circ \delta) \# (\eta' \circ \eta)(A) = (\delta' \circ \delta)F_3(A) \cdot G_1(\eta' \circ \eta)(A) = \delta' F_3(A) \cdot [\delta F_3(A) \cdot G_1\eta'(A)] \cdot G_1\eta(A) = [\delta' F_3(A) \cdot G_2\eta'(A)] \cdot [\delta F_2(A) \cdot G_1\eta(A)] = (\delta' \# \eta')(A) \cdot (\delta \# \eta)(A) = [(\delta' \# \eta') \circ (\delta \# \eta)](A)$. (See Figure 5.4.)

5.2 EQUIVALENCE OF CATEGORIES AND SKELETONS

Just as isomorphisms give a form of equality for objects in a single category, we now consider what shall be meant by "essentially equal" categories. One obvious notion is that of "isomorphism": the categories A and B are isomorphic, with F an isomorphism from A to B, if there are functors $F: A \to B$ and $G: B \to A$ such that $G \cdot F = 1_A$ and $F \cdot G = 1_B$.

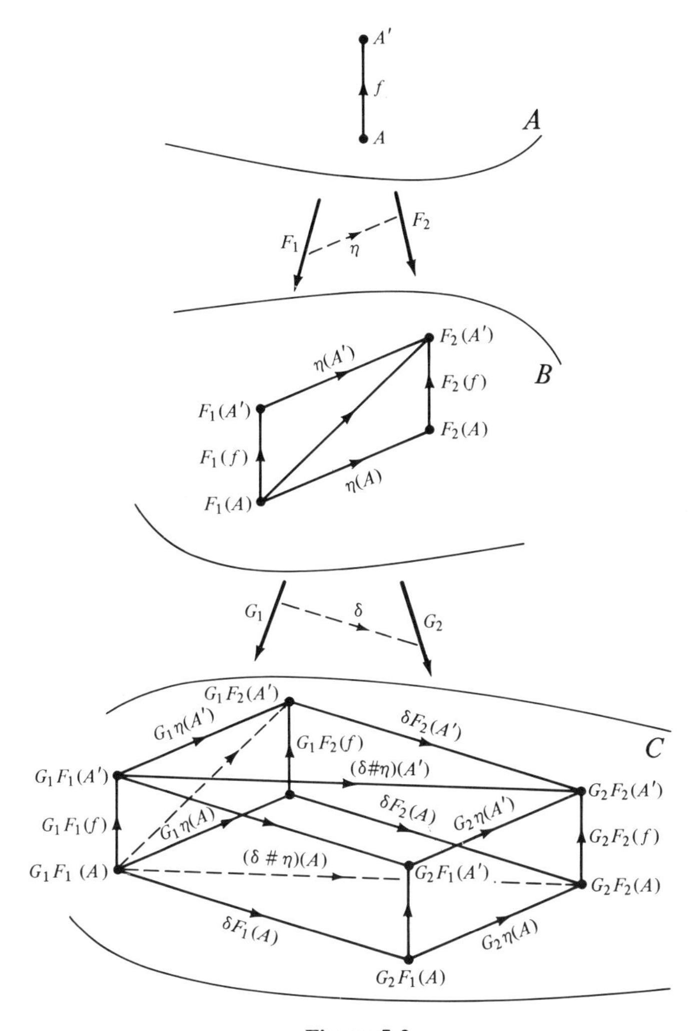

Figure 5.2

This is a strong relation; a weaker relation is often true and still gives near equality of categories. This we now define:

Definition 5.2.1 The categories A and B are called *equivalent*, and $(\eta, \nu; F, G)$ is called an *equivalence scheme* for (A, B), if there exist functors F: $A \rightarrow B$, G: $B \rightarrow A$, and natural equivalences η: $1_B \rightarrow (F \cdot G)$, and ν: $(G \cdot F) \rightarrow 1_A$.

This relation between categories can be visualized by using special subcategories of A, B called *skeletons* in them. So we define:

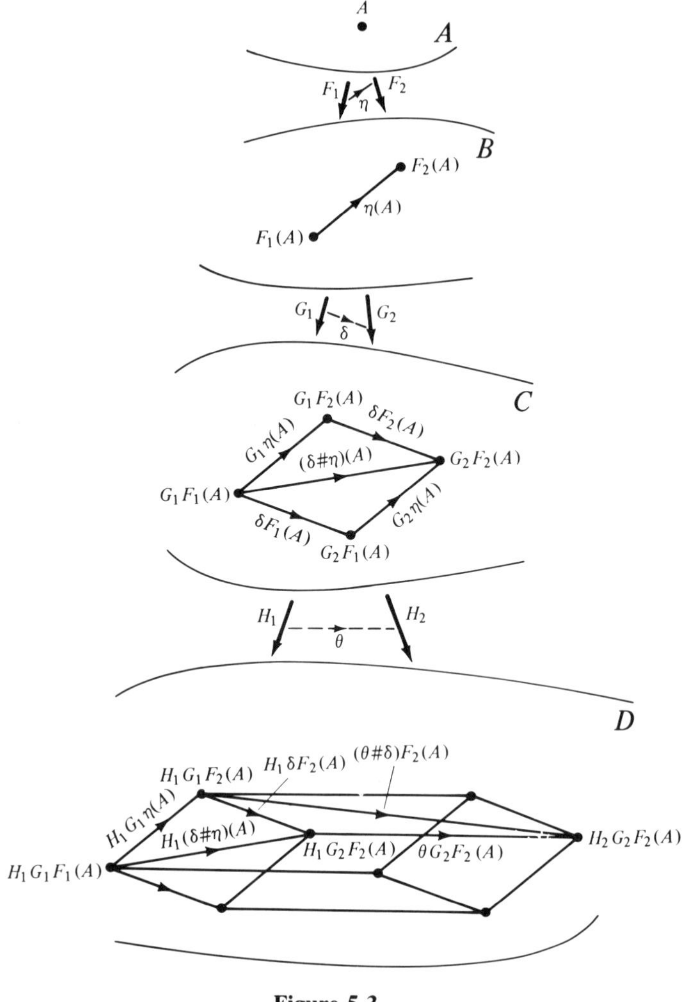

Figure 5.3

Definition 5.2.2 A category C is called a *skeletal category* if each isomorphism in C is an identity morphism [of the form 1_C for a C of $O(C)$].

Definition 5.2.3 We call a full subcategory C of a category a *skeleton* of A if (i) C is a skeletal category and (ii) each object of A is isomorphic with some object of C.

For instance, in the category S, a skeleton is formed by the well-ordered sets of ordinals (assuming the well-ordering hypothesis).

The connection between these notions is examined next.

Theorem 5.2.1 *(a) Isomorphic categories are equivalent; and equivalent skeletal categories are isomorphic.*

(b) For a category A and a skeleton C of A, there is an equivalence scheme $(\eta, \nu; S, I)$ with $I: C \to A$ the inclusion functor, which is dense and faithful, and $S: A \to C$ a functor that is full, faithful, and dense; $S \cdot I = 1_C$.

(c) Two categories are equivalent if and only if a skeleton of one is isomorphic with a skeleton of the other. (Then each skeleton of either is isomorphic with every skeleton of the other.)

(d) Given a faithful, full, and dense functor $F: A \to B$, there is an equivalence scheme $(\eta, \nu; F, G)$ for (A, B) with the further property that $(1_F \# \nu)^r = (\eta \# 1_F)$ and $(\nu \# 1_G)^r = (1_G \# \eta)$, meaning thereby that for any object A *of A, $(1_F \# \nu)(\mathrm{A})$ and $(\eta \# 1_F)(\mathrm{A})$ are inverse isomorphisms, and for any object* B *of B, $(\nu \# 1_G)(\mathrm{B})$ and $(1_G \# \eta)(\mathrm{B})$ are inverse isomorphisms.*

PROOF. (a) It is clear from the definitions that an isomorphism is an equivalence; and if $(\eta, \nu; F, G)$ is an equivalence scheme for the skeletal categories (A, B), then $\eta(\mathrm{B}): \mathrm{B} \to FG(\mathrm{B})$, being an isomorphism in a

Figure 5.4

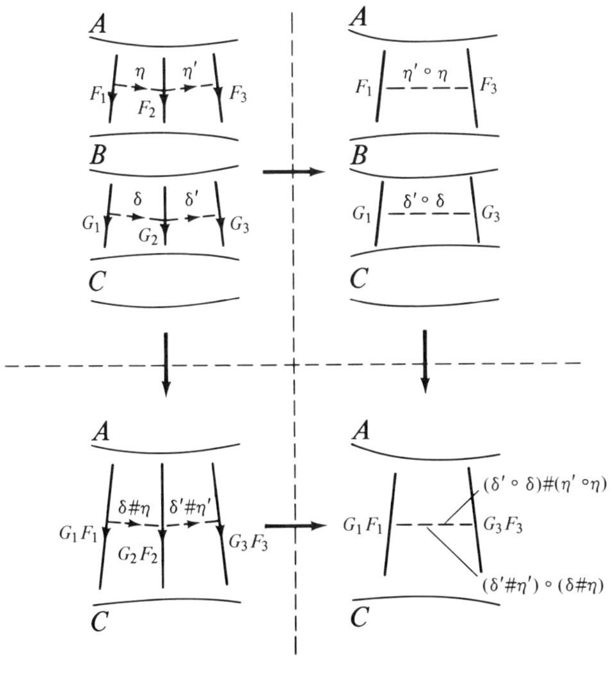

skeletal category, must be 1_B. Thus $FG = 1_B$; similarly, $GF = 1_A$. Or (F, G) gives an isomorphism of the categories (A, B).

(b) If C is a skeleton of A and I denotes the inclusion functor, it is surely a faithful functor; it is dense, since each object of A is known to be isomorphic to an object of C. Further, since C is skeletal, each object A of A determines a unique object of C, which we shall denote by $S(\mathrm{A})$, that is isomorphic with A. For A from $O(C)$ it is clear that $S(\mathrm{A})$ must be the same as A. We shall also choose, for each A from $O(A)$, a specific isomorphism $j(\mathrm{A})$ from A to $S(\mathrm{A})$. Then we can define the functor S from A to C by: $S(\mathrm{A})$ is the object of C isomorphic with A, for an object A of A; and $S(f\colon \mathrm{A} \to \mathrm{A}') = j(\mathrm{A}') \cdot f \cdot [j(\mathrm{A})]^r$ for a typical morphism f from A. It is quite easy to verify that this S is indeed a functor from A to C; that it is faithful, full, and dense; and that $S \cdot I = 1_C$. We set $\eta(\mathrm{C}) = 1_{\mathrm{C}}\colon \mathrm{C} \to SI(\mathrm{C})$ for each C from $O(C)$, and $\nu(\mathrm{A}) = [j(\mathrm{A})]^r\colon IS(\mathrm{A}) \to \mathrm{A}$ for each A from $O(A)$. Evidently η is a natural transformation from 1_C to SI, and a natural equivalence, too, since each $\eta(C)$ is an isomorphism in C. For an $f\colon \mathrm{A} \to \mathrm{A}'$ from $M(A)$, $IS(f) = I(j(\mathrm{A}') \cdot f \cdot [j(\mathrm{A})]^r) = j(\mathrm{A}') \cdot f \cdot [j(\mathrm{A})]^r$, from which we get $[j(\mathrm{A}')]^r \cdot IS(f) = f \cdot [j(\mathrm{A})]^r$ or $\nu(\mathrm{A}') \cdot IS(f) = f \cdot \nu(\mathrm{A})$, showing that ν is also a natural transformation from IS to 1_A; again this is a natural equivalence, since $\nu(\mathrm{A})$ is an isomorphism in A for each A of $O(A)$. Thus we do have an equivalence scheme $(\eta, \nu; S, I)$ for (A, C). Note also that for any A of $O(A)$, we have $S(\nu(\mathrm{A})) = j(\mathrm{A}) \cdot \nu(\mathrm{A}) \cdot [j(IS(\mathrm{A}))]^r = j(\mathrm{A}) \cdot [j(\mathrm{A})]^r \cdot 1_{S(\mathrm{A})}$ [since $IS(\mathrm{A}) = S(\mathrm{A})$ is in $O(C)$] $= 1_{S(\mathrm{A})}$.

(c) We prove this in two parts: (c1) We assume that skeletons C, C' are chosen from the given categories A, A', respectively, and that there is an isomorphism (F, G) for (C, C'). Then we shall show that there is an equivalence scheme $(\eta^*, \nu^*; F^*, G^*)$ for (A, A'). For the skeleton C of A we associate, as in part (b), an equivalence scheme $(\eta, \nu; S, I)$ and similarly one $(\eta', \nu'; S', I')$ for (A', C'). From our hypothesis, we have $G \cdot F = 1_C$ and $F \cdot G = 1_{C'}$. We now define $F^* = I'FS$ and $G^* = IGS'$. Being compositions of functors, these are functors $F^*\colon A \to A'$ and $G^*\colon A' \to A$. We have further $G^*F^* = IG[S'I']FS = IG[1_{C'}F]S = I[GF]S = [I1_C]S = IS$; similarly, we get $F^*G^* = I'S'$. Using the natural equivalences ν, ν', we set $\nu^*(\mathrm{A}) = \nu(\mathrm{A})\colon G^*F^*(\mathrm{A}) = IS(\mathrm{A}) \to \mathrm{A}$ for any A of $O(A)$, and $\eta^*(\mathrm{A}') = [\nu'(\mathrm{A}')]^r\colon \mathrm{A}' \to I'S'(\mathrm{A}') = G^*F^*(\mathrm{A}')$ for each A′ of $O(A')$. These ν^*, η^* are natural equivalences too, and provide the equivalence scheme $(\eta^*, \nu^*; F^*, G^*)$ for the pair (A, A').

(c2) Assume now that the equivalence scheme $(\eta^*, \nu^*; F^*, G^*)$ for (A, A') is given; we choose any skeletons C, C' from A, A', and use the equivalence schemes $(\eta, \nu; S, I)$ and $(\eta', \nu'; S', I')$ for (A, C) and (A', C') as in (c1). We shall prove that there is an isomorphism (F, G) for (C, C'). Set $F = S'F^*I$ and $G = SG^*I'$. These are surely functors from C to C'

and vice versa. To prove then that $FG = 1_{C'}$ and $GF = 1_C$, we shall prove the first, the second being proved similarly. (See Figure 5.5.)

Let f: C → D be a typical morphism from C. Since $\nu^*: G^*F^* \to 1_A$ is a natural equivalence, $\nu^*(I(\mathrm{C}))$, $\nu^*(I(\mathrm{D}))$ are isomorphisms such that

$$I(f) \cdot \nu^*(I(\mathrm{C})) = \nu^*(I(\mathrm{D})) \cdot G^*F^*I(f). \tag{5.2.1}$$

Since ν': $I'S' \to 1_{A'}$ is a natural equivalence, $\nu'(F^*I(\mathrm{C}))$, $\nu'(F^*I(\mathrm{D}))$ are isomorphisms in A' such that

$$F^*I(f) \cdot \nu'(F(I(\mathrm{C})) = \nu'(F^*I(\mathrm{D})) \cdot I'S'F^*I(f), \tag{5.2.2}$$

and then the functor G^* gives isomorphisms $G^*\nu'(F^*I(C))$, $G^*\nu'(F^*I(D))$ in A such that

$$G^*F^*I(f) \cdot G^*\nu'(F^*I(\mathrm{C})) = G^*\nu'(F^*I(\mathrm{D})) \cdot G^*I'S'F^*I(f). \tag{5.2.3}$$

Setting $k(\mathrm{C}) = \nu^*(I(\mathrm{C})) \cdot G^*\nu'(F^*I(\mathrm{C}))$ for C of $O(C)$, and $k(\mathrm{D})$ for D of $O(C)$ similarly, we see that they are isomorphisms in A' such that, by (5.2.1) and (5.2.3), $I(f) \cdot k(\mathrm{C}) = k(\mathrm{D}) \cdot G^*I'S'F^*I(f)$. Applying S to this, and remembering that the images $Sk(\mathrm{C})$, $Sk(\mathrm{D})$, being isomorphisms in C, have to be identities, we get $SI(f) = SG^*I'S'F^*I(f) = GF(f)$; but

Figure 5.5

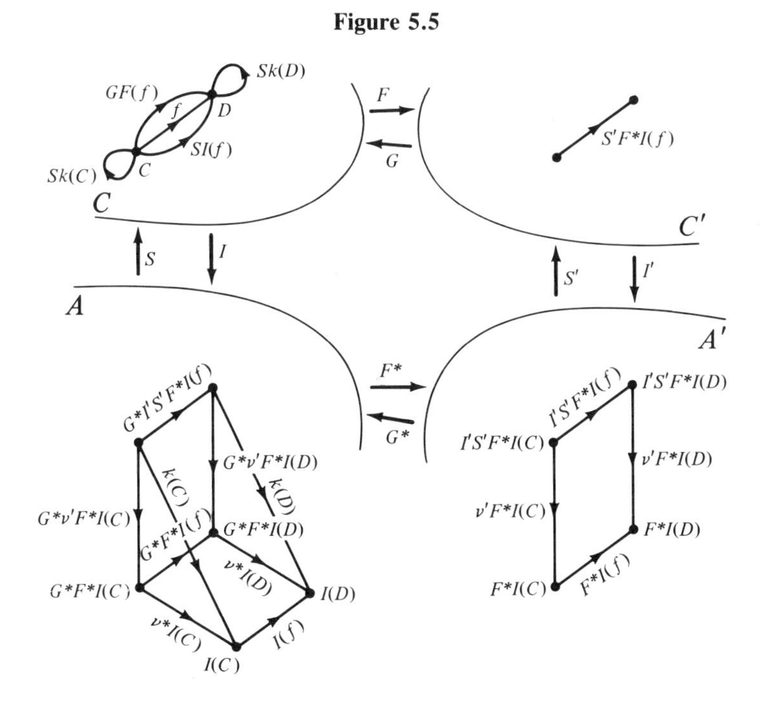

for f from $M(C)$, $SI(f)$ is surely the same as f. Hence finally we have $GF(f) = f$ for all f in $M(C)$, proving that $GF = 1_C$.

(d) Let us now assume that we are given a full, faithful, and dense functor $F: A \to B$. We choose a skeleton C in A and have an associated equivalence scheme $(\eta', \nu'; S', I')$ for (A, C), as in (b). The functor I' is also full, dense, and faithful; hence, so is the composite $FI': C \to B$. The class of morphisms $FI'(M(C))$ is closed for compositions [in $M(B)$] and, with the class $FI'(O(C))$ for its objects, forms a full subcategory that is also dense in B; we denote this by D. [If $FI'(g) \cdot FI'(h)$ is defined in B, where h, g are from $\hom_C(\mathrm{C}_1, \mathrm{C}_2)$ and $\hom_C(\mathrm{C}_2', \mathrm{C}_3)$, then $FI'(\mathrm{C}_2) = FI'(\mathrm{C}_2') = \mathrm{D}_2$ is the domain of $FI'(g)$ and codomain of $FI'(h)$. The identity 1_{D_2} then is an isomorphism that must be the image of an isomorphism k in $\hom_C(\mathrm{C}_2, \mathrm{C}_2')$. Then $g \cdot k \cdot h$ is in $\hom_C(\mathrm{C}_1, \mathrm{C}_3)$ and $FI'(g \cdot k \cdot h) = FI'(g) \cdot FI'(h)$. Thus the composite belongs to $M(D) = FI'(M(C))$.] The rest is easy, to see that D is indeed a full and dense subcategory of B. Finally, we prove that D, like C, is a skeletal category, hence a skeleton of B. For any isomorphism j in D must be the image of an isomorphism k in C, since a faithful, full, and dense functor reflects isomorphisms; but then k must be an identity in C, because C is skeletal. Then $j = FI'(k)$ is also an identity in D. Thus D is skeletal. Let us then fix an equivalence scheme $(\eta'', \nu''; S'', I'')$ for (B, D). If we set $F' = S'' \cdot F \cdot I': C \to D$, it is clear that F' is an isomorphism between the skeletal categories C, D with $I''F' = FI'$. Hence this has an inverse $G': D \to C$ with $F'G' = 1_D$, $G'F' = 1_C$. We finally define $G: B \to A$ by $G = I'G'S''$. (See Figure 5.6.)

We have then $FG = [FI']G'S'' = I''[F'G']S'' = [I''1_D]S'' = I''S''$. We can now define for any B from $O(B)$, $\eta(\mathrm{B}): \mathrm{B} \to FG(\mathrm{B}) = I'' \cdot S''(\mathrm{B})$ to be the same as $[\nu''(\mathrm{B})]^r$. Since ν'' is a natural equivalence from $I'' \cdot S''$ to 1_B, η is a natural equivalence from 1_B to FG. We next compute FGF; $FGF = FI'G'S''F = I''F'G'S''F = I''1_DS''F = I'' \cdot S'' \cdot F$. Hence, for any A from $O(A)$, $F(GF(\mathrm{A})) = (I'' \cdot S'')(F(\mathrm{A}))$. Since $\nu''(F(\mathrm{A})): I'' \cdot S''(F(\mathrm{A})) \to F(\mathrm{A})$ is an isomorphism in B, it reflects (by the full, faithful, dense functor F) to an isomorphism $h: GF(\mathrm{A}) \to \mathrm{A}$ [we are using the fact that $I'' \cdot S''(F(\mathrm{A})) = F(GF(\mathrm{A}))$]. We denote this h by $\nu(\mathrm{A})$. Then $\nu(\mathrm{A})$ is the unique isomorphism $h: GF(\mathrm{A}) \to \mathrm{A}$ in A for which $F(h) = \nu''(F(\mathrm{A}))$. We check now that ν is a natural equivalence from GF to 1_A; since $\nu(\mathrm{A})$ is always an isomorphism in A, it is enough to show that ν is a natural transformation from GF to 1_A. If $f: \mathrm{A} \to \mathrm{A}'$ is any morphism from A, the natural transformation ν'' from $I'' \cdot S''$ to 1_B gives, for the morphism $F(f): F(\mathrm{A}) \to F(\mathrm{A}')$, the commutativity relation $F(f) \cdot \nu''(F(\mathrm{A})) = \nu''(F(\mathrm{A}')) \cdot I'' \cdot S''(F(f)) = \nu''(F(\mathrm{A}')) \cdot FGF(f)$. That is, the faithful functor F takes $f \cdot \nu(\mathrm{A})$ and $\nu(\mathrm{A}') \cdot GF(f)$ to the same morphism in B; hence they should also be equal. But this is the commutativity to be proved in order for ν to be a natural transformation.

Finally, for any A from $O(A)$, $\eta(F(\mathrm{A})) = [\nu''(F(\mathrm{A}))]^r = [F(\nu(\mathrm{A}))]^r$, so

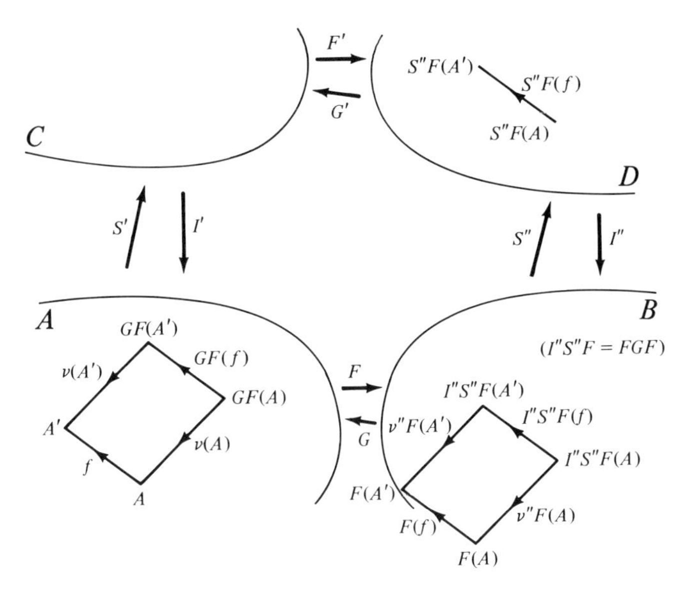

Figure 5.6

that $(\eta \# 1_F)^r = (1_F \# \nu)$, while for any B from $O(B)$, the natural transformation ν'': $FG = I'' \cdot S'' \rightarrow 1_B$ gives for the morphism $\nu''(\mathrm{B})$: $FG(\mathrm{B}) \rightarrow \mathrm{B}$ a commutative relation $\nu''(\mathrm{B}) \cdot \nu''(FG(\mathrm{B})) = \nu''(\mathrm{B}) \cdot (FG)(\nu''(\mathrm{B}))$. Since $\nu''(\mathrm{B})$ is an isomorphism, it is left-cancelative; so we get $\nu''(FG(\mathrm{B})) = (FG(\nu''(\mathrm{B}))$. The definition of ν implies that $\nu(G(\mathrm{B})) = G(\nu''(\mathrm{B})) = G[\eta(\mathrm{B})]^r = [G(\eta(\mathrm{B}))]^r$; this gives then $(1_G \# \eta)^r = (\nu \# 1_G)$. □

When A is a full, dense subcategory of a category B, the inclusion functor from A to B being full, dense, and faithful, the category A is equivalent to B. Since this relation of equivalence of categories is symmetric and transitive, and includes isomorphisms, any two categories A, A' that are isomorphic, respectively, to two full dense subcategories of some category B would be equivalent too. This is really an alternative description of equivalence, for it can be shown (see the exercises at the end of the chapter for details) that two categories are equivalent if and only if they are isomorphs of two subcategories that are full and dense in some suitable category B.

When B is a category and A^* is a subclass of $O(B)$ such that each object of B is isomorphic with some object in A^* [we call such an A^* a dense subclass of $O(B)$], then we can form the full subcategory A of B that has A^* for $O(A)$ [by setting $\hom_A(\mathrm{A}_1, \mathrm{A}_2) = \hom_B(\mathrm{A}_1, \mathrm{A}_2)$ for any A_1, A_2 from A^*]; this subcategory A is also a dense subcategory; so A is equivalent to B. This type of equivalence often occurs, as we see in what follows.

Example 1. From Theorem 1.2.2 we know that in the categories Sgp^*, $RSgp^*$, $ISgp$, and Gp of the semigroups with unit, regular semigroups with unit, inverse semigroups, and groups (with groupoid homomorphisms for the first and regular groupoid homomorphisms for the later ones, as morphisms) there are dense subclasses formed by semigroups of mapping relations on a set, semigroups of mapping relations on a set with a suitable choice of regular inverse for each relation, inverse semigroups of partial bijections on a set, and groups of bijective mappings in a set, respectively; so we have here typical dense full subcategories that are equivalent to the originals, but consist in each case of elements that are relations whose composition is relational multiplication. In the same way the corollary to Lemma 1.3.2 shows that the category FA (of F-algebras and their F-homomorphisms) has a dense subclass consisting of "quotients of free F-algebras," and so these generate a full subcategory equivalent to FA.

Example 2. If NS, CS, and T denote the categories of neighborhood spaces, convergence spaces, and topological spaces, respectively, each with the continuous mappings as morphisms, then from Lemma 1.4.1 we see that there are functors I: $NS \to CS$, J: $CS \to T$, and K: $T \to NS$ such that $K(JI)$ is naturally equivalent to 1_{NS}, $JIK = 1_T$, and IKJ is naturally equivalent to 1_{CS}. Thus these three categories are equivalent in pairs. If we take the subcategory FNS of NS having for objects neighborhood spaces with each $N(x)$ (the set of neighborhoods at x) a filter over the basic set X, then FNS and T can be shown to be isomorphic.

Example 3. In the category SU of semiuniform spaces and their uniform maps, let $SMSU$ and ESU denote those full subcategories whose objects are specified by the conditions (a) they are semimetrizable semiuniform spaces, or (b) they are semiuniform spaces with an enumerable indexing set J; then from Theorem 1.5.1 we can deduce that these two subcategories are equivalent. The same theorem also gives the equivalence of SUT and T, QUT and CR, and UT and (CR, T_0) where these are all full subcategories of T. (SUT, QUT, and UT denote the subcategories whose objects are spaces admitting a compatible semiuniformity, quasiuniformity, and uniformity, respectively.)

Example 4. When A^{op} and B are equivalent categories, we say that A, B are *dually equivalent* categories. Many classical duality theorems give just such pairs:

VSF, the category of finite-dimensional vector spaces over a field F, is *self-dual*, that is, dually equivalent to itself.

CAG and AG, the categories of compact Abelian groups and of Abelian groups, respectively, are dually equivalent (or, as we also say, *dual categories*); this is the essence of Pontrjagin's duality theory.

LCAG, the category of locally compact Abelian groups, is self-dual; this is the extension of Pontrjagin's duality.

CBA and *BS*, the categories fo complete Boolean algebras and of Boolean spaces, respectively, are dual categories; this is the essence of Stone's duality.

5.3 FUNCTOR CATEGORIES

Given two fixed categories A and J, the functors from J to A can be thought of as objects, and natural transformations between such functors can be treated as morphisms in a form of *metacategory*. This is itself a category when the category J is small; that is, $O(J)$ and $M(J)$ are both sets. We then define A^J, or $F(J \to A)$, to be a category with $O(F(J \to A))$ = [the class of functors $F\colon J \to A$], $M(F(J \to A))$ = [natural transformations between pairs of functors from J to A], setting, for F, G from $O(F(J \to A))$, $\hom_{F(J\to A)}(F, G)$ = [the set of natural transformations $\eta\colon F \to G$]. This last is a set, since it is a subclass of the power set $P[\cup \{\hom_A(F(j) \to G(j)\colon j \text{ in } O(J) = \text{a set}\}]$. The composition of an $\eta\colon F \to G$ followed by a $\delta\colon G \to H$ is $\delta \circ \eta$; and there is a $1_F\colon F \to F$ for each F of $O(F(J \to A))$. We consider some of these functor categories now.

1. Given the categories A, J (with J a small category), we have a *constant functor* $c^*\colon A \to A^J$; for an object A of A, $c^*(\mathrm{A})(j) = \mathrm{A}$ for each j of $O(J)$; and for an f from $M(J)$, $c^*(\mathrm{A})(f) = 1_{\mathrm{A}}$. For a morphism $g\colon \mathrm{A} \to \mathrm{A}'$ in A, $c^*(g)(j) = g\colon c^*(\mathrm{A})(j) \to c^*(\mathrm{A}')(j)$ for each j of $O(J)$. We can check that this gives a functor from A to $F(J \to A)$.
2. For a category A and small category J we have a functor $E\colon J \times A^J \to A$, called the *evaluation functor*, defined as follows. For an object (j, F) from $J \times F(J \to A)$, $E(j, F)$ is the object $F(j)$ in A; and for a morphism (g, η) from (j, F) to (k, G) in $J \times F(J \to A)$, $E(g, \eta)$ is the morphism $\eta(k) \cdot F(g) = G(g) \cdot \eta(j)$. Again the verification that this does indeed give a functor, as claimed, is not difficult.
3. A morphism η of $F(J \to A)$ would be an isomorphism if each $\eta(j)\colon j$ in $O(J)$ were an isomorphism in A; similarly, we can show that η would be a monomorphism, epimorphism, or bimorphism, a constant or coconstant or zero morphism, if each $\eta(j)$ is such in A. (See the exercises at the end of this chapter.)

Keeping the small category J fixed, we get an association $A \to F(J \to A)$ between categories. This can be extended to cover functors between categories also: to an $H\colon A \to B$, we can associate an $H_*\colon F(J \to A) \to F(J \to B)$ by setting $H_*(F) = H \cdot F$, and $H_*(\eta\colon F \to G) = (\eta \mathbin{\#} 1_H)\colon H \cdot F \to H \cdot G$, for any F of $O(F(J \to A))$ and any η of $M(F(J \to A))$.

4. When X is an initial, final, or zero object for A, then $c^*(X)$ is an initial, final, or zero object for $F(J \to A)$. If in A there is a constant and a coconstant morphism between each pair of objects, the same is true in $F(J \to A)$. Hence, by theorem 3.4.1, $F(J \to A)$ would be a pointed category when A is such.

Keeping the category A fixed, for variable J we have a passage $J \to F(J \to A)$ between categories. This can be extended to functors: For H: $K \to J$, a functor between small categories, we define H^*: $F(J \to A) \to F(K \to A)$ as follows. For an F_1 in $O[F(J \to A)]$, $H^*(F_1) = F_1 \cdot H$, and for an η in $M[F(J \to A)]$, $H^*[\eta: F_1 \to F_2] = (1_H \# \eta): F_1 \cdot H \to F_2 \cdot H$.

5.4 NATURAL TRANSFORMATIONS FOR FEEBLE FUNCTORS

When we wish to get an analogue of the natural transformation for a pair of feeble functors between the same categories, we find we have really to deal with two possible types. We first define these:

Definition 5.4.1 If F and G are both feeble functors from the category A to the category B, we say that η: $F \to G$ is a *left-* [a *right-*] *natural transformation* from F to G if, for each object A of A, $\eta(\mathrm{A})$ is a nonnull family of morphisms in B from $F(\mathrm{A})$ to $G(\mathrm{A})$ such that a morphism f: $\mathrm{A} \to \mathrm{A}'$ in A gives rise to the inequality $\eta(\mathrm{A}') \cdot F(f) \supseteq G(f) \cdot \eta(\mathrm{A})$ [$G(f)\eta(\mathrm{A}) \supseteq \eta(\mathrm{A}') \cdot F(f)$]. Note that the larger of the two composites has the η term on the left, or on the right, according as the natural transformation is a left or a right one. When η is both left and right natural, we call it just a *natural transformation* between the feeble functors.

Clearly, if F, G are functors and η is a natural transformation, we can consider the former a pair of special feeble functors and the latter a natural transformation between them.

The natural transformation η: $F_1 \to F_2$, when F_1, F_2 are feeble functors from A to B, can be viewed as a cofull relation from the class $O(A)$ to the class $M(B)$ satisfying a suitable inclusion condition corresponding to a morphism f in A. We again consider a horizontal and vertical composition of such natural transformations; the main results are in

Lemma 5.4.1 *(a) Given feeble functors F_1, $i = 1, 2, 3$, from A to B, and two left- [right-] natural transformations η: $F_1 \to F_2$, η': $F_2 \to F_3$, we have a left- [right-] natural transformation $(\eta' \circ \eta)$ from F_1 to F_3 defined by setting $(\eta' \circ \eta)(\mathrm{A}) = \eta'(\mathrm{A}) \cdot \eta(\mathrm{A})$ for each A in $O(A)$. This type of horizontal composition of left- [right-] natural transformations is an associative operation.*

(b) Given feeble functors F_1, F_2 *from A to B, and* G_1, G_2 *from B to C, and left- [right-] natural transformations* $\eta: F_1 \to F_2$, $\delta: G_1 \to G_2$, *we define two forms of vertical composition* $(\delta^{\#}\eta)$ *and* $(\delta_{\#}\eta)$, *which are again left- [right-] natural transformations from* (G_1F_1) *to* (G_2F_2), *by setting* $(\delta^{\#}\eta)(\mathrm{A}) = G_2(\eta(\mathrm{A})) \cdot \delta(F_1(\mathrm{A}))$ *and* $(\delta_{\#}\eta)(\mathrm{A}) = \delta(F_2(\mathrm{A})) \cdot G_1(\eta(\mathrm{A}))$ *for each object* A *from A. Each of these is an associative operation.*

(c) When η *and* δ *are left- [right-] natural transformations defined as in (b),* $(\delta^{\#}\eta) \geq (\delta_{\#}\eta)[(\delta^{\#}\eta) \leq (\delta_{\#}\eta)]$; *that is, for each* A *from O(A),* $(\delta^{\#}\eta)(\mathrm{A}) \supseteq$ [*or* $\subseteq$] $(\delta_{\#}\eta)(\mathrm{A})$.

(d) When there are three feeble functors F_i *from A to B and three* G_i *from B to C (i = 1, 2, 3), and left- [right-] natural transformations* $\eta: F_1 \to F_2$, $\eta': F_2 \to F_3$, $\delta: G_1 \to G_2$, *and* $\delta': G_2 \to G_3$, *we have the following "interchange" inequalities:* $(\delta' \circ \delta)^{\#}(\eta' \circ \eta) \geq (\delta'^{\#}\eta') \circ (\delta^{\#}\eta)$, $(\delta' \circ \delta)_{\#}(\eta' \circ \eta) \leq (\delta_{\#}'\eta') \circ (\delta_{\#}\eta)$ [*the same with* $\leq$, $\geq$ *interchanged*].

PROOF. (a) To check that $(\eta' \circ \eta)$ is left natural, when η', η are such: For any morphism f: $\mathrm{A} \to \mathrm{A}'$ in A, we have $(\eta' \circ \eta)(\mathrm{A}') \cdot F_1(f) = \eta'(\mathrm{A}') \cdot \eta(\mathrm{A}') \cdot F_1(f) \supseteq \eta'(\mathrm{A}') \cdot F_2(f) \cdot \eta(\mathrm{A}) \supseteq F_3(f) \cdot \eta'(\mathrm{A}) \cdot \eta(\mathrm{A}) = F_3(f) \cdot (\eta' \circ \eta)(\mathrm{A})$, which is what ensures that $(\eta' \circ \eta)$ is a left-natural transformation. A similar proof works for the composition of right-natural transformations. Associativity is easily proved: if η'': $F_3 \to F_4$ is a third [left- right-] natural transformation, besides η, η' as above, then a typical morphism (of B) from $[(\eta'' \circ \eta') \circ \eta](\mathrm{A})$ is of the form $(p \cdot q) \cdot r$, where p, q, r are from $\eta''(\mathrm{A})$, $\eta'(\mathrm{A})$, and $\eta(\mathrm{A})$, respectively; but then this $(p \cdot q) \cdot r = p \cdot (q \cdot r)$ also belongs to $[\eta'' \circ (\eta' \circ \eta)](\mathrm{A})$, and conversely.

(b) Assuming that η: $F_1 \to F_2$ and δ: $G_1 \to G_2$ are left-natural transformations, we check that the composites defined here are also such; given a morphism f: $\mathrm{A} \to \mathrm{A}'$ from A, $(\delta^{\#}\eta)(\mathrm{A}') \cdot G_1F_1(f) = G_2(\eta(\mathrm{A}')) \cdot [\delta(F_1(\mathrm{A}')) \cdot G_1F_1(f)] \supseteq [G_2(\eta(\mathrm{A}')) \cdot G_2F_1(f)] \cdot \delta(F_1(\mathrm{A})) \supseteq G_2F_2(f) \cdot [G_2(\eta(\mathrm{A})) \cdot \delta(F_1(\mathrm{A}))] = G_2F_2(f) \cdot (\delta^{\#}\eta)(\mathrm{A})$. This proves that $(\delta^{\#}\eta)$ is a left-natural transformation. Similarly, $(\delta_{\#}\eta)(\mathrm{A}') \cdot G_1F_1(f) = \delta(F_2(\mathrm{A}')) \cdot [G_1(\eta(\mathrm{A}')) \cdot G_1F_1(f)] \supseteq [\delta(F_2(\mathrm{A}')) \cdot G_1F_2(f)] \cdot G_1(\eta(\mathrm{A})) \supseteq G_2F_2(f) \cdot [\delta(F_2(\mathrm{A})) \cdot G_1(\eta(\mathrm{A}))] = G_2F_2(f) \cdot (\delta_{\#}\eta)(\mathrm{A})$, proving that $(\delta_{\#}\eta)$ is a left-natural transformation. In a similar manner we can prove that when both η and δ are right-natural transformations, then the composites $(\delta^{\#}\eta)$ and $(\delta_{\#}\eta)$ are also such. (See Figure 5.7.)

To consider associativity of these operations, let H_1, H_2 be functors from C to D, with a left- [right-] natural transformation θ: $H_1 \to H_2$. For each associativity we check the effect of the triple products on a typical object A of A; thus

$$\begin{aligned}[(\theta^{\#}\delta)^{\#}\eta](\mathrm{A}) &= H_2G_2(\eta(\mathrm{A})) \cdot [(\theta^{\#}\delta)(F_1(\mathrm{A}))] \\ &= [H_2G_2(\eta(\mathrm{A})) \cdot H_2\delta(F_1(\mathrm{A}))] \cdot \theta(G_1F_1(\mathrm{A})) \\ &= [\theta^{\#}(\delta^{\#}\eta)](\mathrm{A});\end{aligned}$$

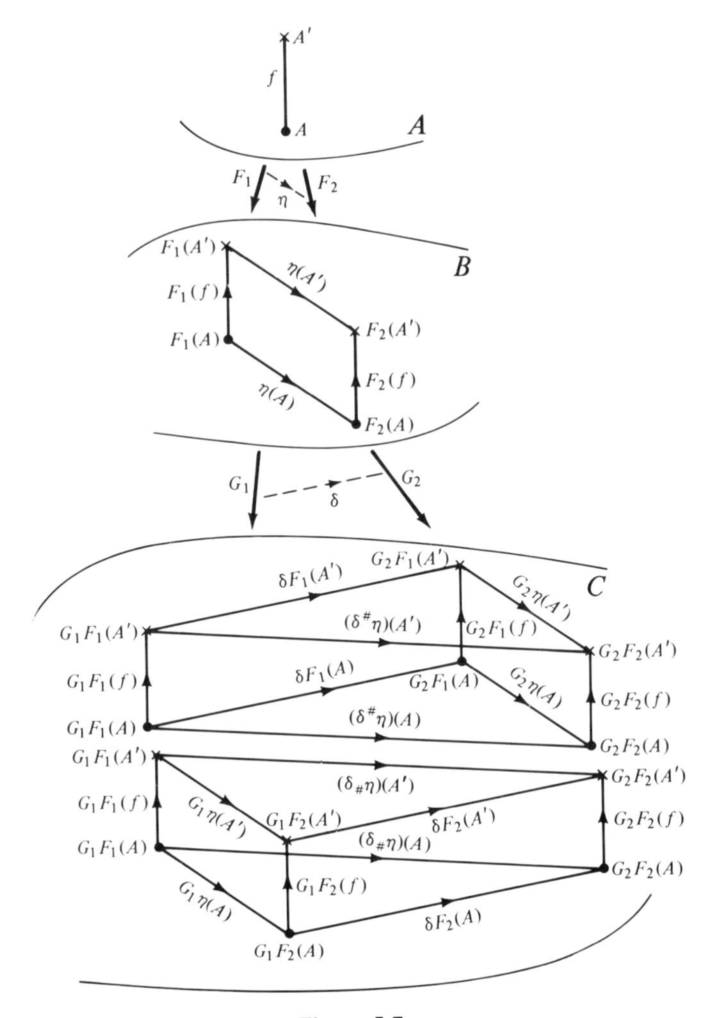

Figure 5.7

and similarly,

$$
\begin{aligned}
[(\theta_{\#}\delta)_{\#}\eta](A) &= [(\theta_{\#}\delta)(F_2(A))] \cdot H_1G_1(\eta(A)) \\
&= \theta(G_2F_2(A)) \cdot [H_1(\delta(F_2(A))) \cdot H_1G_1(\eta(A))] \\
&= \theta(G_2F_2(A)) \cdot H_1[(\delta_{\#}\eta)(A)] \\
&= [\theta_{\#}(\delta_{\#}\eta)](A).
\end{aligned}
$$

(c) Proofs of these relations follow immediately from the definitions.

(d)

$$
\begin{aligned}
[(\delta' \circ \delta)^{\#}(\eta' \circ \eta)](\mathrm{A}) &= G_3((\eta' \circ \eta)(\mathrm{A})) \cdot (\delta' \circ \delta)(F_1(\mathrm{A})) \\
&= G_3(\eta'(\mathrm{A})) \cdot G_3(\eta(\mathrm{A})) \cdot \delta'(F_1(\mathrm{A})) \cdot \delta(F_1(\mathrm{A})) \\
&\supseteq G_3(\eta'(\mathrm{A})) \cdot \delta'(F_2(\mathrm{A})) \cdot G_2(\eta(\mathrm{A})) \cdot \delta(F_1(\mathrm{A})) \\
&= (\delta'^{\#}\eta')(\mathrm{A}) \cdot (\delta^{\#}\eta)(\mathrm{A}) = [(\delta'^{\#}\eta') \circ (\delta^{\#}\eta)](\mathrm{A}) \\
[(\delta' \circ \delta)_{\#}(\eta' \circ \eta)](\mathrm{A}) &= (\delta' \circ \delta)(F_3(\mathrm{A})) \cdot G_1[(\eta' \circ \eta)(\mathrm{A})] \\
&= \delta'(F_3(\mathrm{A})) \cdot \delta(\mathrm{F}_3(\mathrm{A})) \cdot G_1(\eta'(\mathrm{A})) \cdot G_1(\eta(\mathrm{A})) \\
&\subseteq \delta'(F_3(\mathrm{A})) \cdot G_2(\eta'(\mathrm{A})) \cdot \delta(F_2(\mathrm{A})) \cdot G_1(\eta(\mathrm{A})) \\
&= (\delta'_{\#}\eta')(\mathrm{A}) \cdot (\delta_{\#}\eta)(\mathrm{A}) = [(\delta'_{\#}\eta') \circ (\delta_{\#}\eta)](\mathrm{A}).
\end{aligned}
$$

These are when the natural transformations are all left ones; for right ones, the inequalities would be reversed. □

EXERCISES AND REMARKS ON CHAPTER 5

1. Show that a morphism η in $F(J \to A)$ is a mono-, epi-, or bimorphism if $\eta(j)$ is such in A for each j of $O(J)$.

We check the case of the epimorphism: if η: $F \to G$, and δ: $G \to H$, θ: $G \to H$ are such that $\delta \circ \eta = \theta \circ \eta$, and if each $\eta(j)$ is epimorphic in A, then for each j of $O(J)$, from $\delta(j) \cdot \eta(j) = \theta(j) \cdot \eta(j)$, we get $\delta(j) = \theta(j)$; that is, $\delta = \theta$; hence η is also an epimorphism.

2. Show that a morphism η in $F(J \to A)$ is a constant, coconstant, or zero morphism in $F(J \to A)$ if, for each j in $O(J)$, $\eta(j)$ is such in A.

We can check for the coconstant case: given η, δ, and θ as in the last exercise, if each $\eta(j)$ is a coconstant morphism, $(\delta \circ \eta)(j) = (\theta \circ \eta)(j)$ for each j; hence, $\delta \circ \eta = \theta \circ \eta$. That is, η is a coconstant morphism.

3. Show that $F(J \to A)$ is a pointed category when A is such.

When A is a pointed category, for each pair of objects A, B from A there is a unique zero morphism $z(\mathrm{A}, \mathrm{B})$ in A. Setting, for any F, G from $O(F(J \to A))$, $\eta(j) = z(F(j), G(j))$, we see, as in the last exercise, that η is both a constant and a coconstant morphism, or a zero morphism, from F to G. Hence, each pair of objects F, G in $F(J \to A)$ is connected by a zero morphism. This implies that $F(J \to A)$ is pointed.

CHAPTER SIX

LIMITS, COLIMITS, COMPLETENESS, COCOMPLETENESS

6.1 PREDECESSORS AND LIMITS OF A FUNCTOR

The notion of limit includes as special cases such constructions as a direct product of a set of objects, an equalizer of a pair of morphisms, and other notions that we shall be using hereafter.

We start with the basic definitions and follow with typical examples.

Definition 6.1.1 Given a functor $F: J \to A$, where J is a small category, a pair (B, p) consisting of an object B of A and a natural transformation $p: c^*(\mathrm{B}) \to F$ is called a *predecessor of F*; (recall that the constant functor $c^*(\mathrm{B})$ takes all objects of J to B and all morphisms of J to 1_B) this means that for each object j of J there is a morphism $p(j): \mathrm{B} \to F(j)$ in A; and for any morphism $g: j \to k$ in J, $F(g) \cdot p(j) = p(k)$.

We can make these predecessors of F the class of objects of a category $Pr(F)$ by setting $\hom_{Pr(F)}[(\mathrm{B}, p), (\mathrm{B}', p')]$ (for a pair of these objects) to be equal to [morphisms m^* of A from B to B′ such that $p'(j) \cdot m^* = p(j)$ for each j of $O(J)$].

Definition 6.1.2 If the category $Pr(F)$ has a final object (B^*, p^*), then we call it a *limit of the functor F in A*.

As a final object it is unique up to isomorphisms in $Pr(F)$, which amounts to saying that B^* is unique up to isomorphs in A.

We consider a few typical and important examples of limits.

6.1.1. The Product

Let J be a small and discrete category [that is, it has a set of objects $O(J)$ and only identity morphisms at these objects as the family of morphisms]; if a functor F: $J \rightarrow A$ has a limit (B*, p^*) in A, it is usual to write B* $= P[F(j)$: j in $O(J)]$ and call this a *product* in A of the indexed set of objects $[F(j)]$ with the *canonical projections* $p^*(j)$. It is clear from our definition of limit that when this happens the projections $p^*(j)$: B* $\rightarrow F(j)$ are morphisms in A, and for any object B of A and family of morphisms $p(j)$: B $\rightarrow F(j)$ [one for each j in $O(J)$], there would be a unique morphism p: B $\rightarrow$ B* in A such that $p(j) = p^*(j) \cdot p$ for each j of J. This unique morphism is denoted by $SP(p_j$: j in $J)$ and is called the *semiproduct* of the family of morphisms $[p(j)]$. Note that the definition of p depends on the choice of the product object B*; if B* is replaced by an isomorph i(B*), where i is an isomorphism, p will be replaced by $i \cdot p$. (See Figure 6.1.)

The most familiar special case of this is when J is a discrete two-element category: $[1, 2] = O(J)$. Then we have the usual direct product $F(1) \times F(2)$ of the objects $F(1)$, $F(2)$ with the usual projection morphisms $p(1)$, $p(2)$ of the product in $F(1)$, $F(2)$. For the concrete categories like FA or T, the product object is usually chosen such that its set base is the usual direct product, or Cartesian product, of the set bases of the two factors. Similar to the usual commutative or associative properties for this binary multiplication, we can formulate commutative and associative laws for the general product, too! (See the exercises at the end of this chapter.)

Figure 6.1

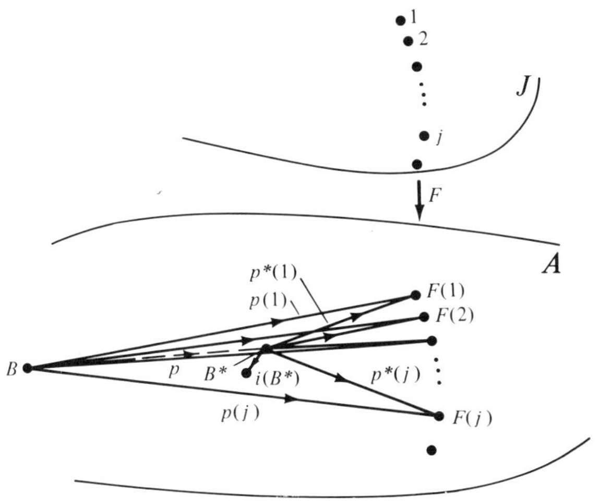

6.1.2. The Equalizer

For this we can take a finite category J with two objects (1, 2) and four morphisms $[1_1, 1_2, f, g]$, where f and g are both from 1 to 2. Any functor $F: J \to A$ then determines a similar set of objects $(F(1), F(2))$ and morphisms $[F(1_1), F(1_2), F(f), F(g)]$, essentially two nontrivial morphisms $F(f)$, $F(g)$ between the same pair of objects $F(1)$, $F(2)$. If there is a limit for this F, such a limit (B^*, p^*) would be characterized by the properties; (a) $\mathrm{B}^* \in O(A)$, $p^*(i)$: $\mathrm{B}^* \to F(i)$, $i = 1, 2$, are morphisms in A, with $F(f) \cdot p^*(1) = p^*(2) = F(g) \cdot p^*(1)$; and (b) when $\mathrm{B} \in O(A)$ and $p(i)$: $\mathrm{B} \to F(i)$, $i = 1, 2$, are morphisms in A such that $F(f) \cdot p(1) = p(2) = F(g) \cdot p(1)$, then there exists a unique morphism p: $\mathrm{B} \to \mathrm{B}^*$ such that $p^*(1) \cdot p = p(1)$ and $p^*(2) \cdot p = p(2)$. If we examine this and compare our earlier definition of an equalizer, we see that (B^*, p^*) is a limit of F iff $(\mathrm{B}^*, p^*(1))$ is an equalizer of the pair of morphisms $[F(f), F(g)]$. (See Figure 6.2.)

6.1.3. The Pullback

In the category A, a pair of morphisms p, q with a common codomain C [say p in $\hom_A(\mathrm{A}, \mathrm{C})$ and q in $\hom_A(\mathrm{B}, \mathrm{C})$] is said to have as a "pullback" the pair of morphisms m, n with common domain D with m in $\hom_A(\mathrm{D}, \mathrm{A})$ and n in $\hom_A(\mathrm{D}, \mathrm{B})$ if $p \cdot m = q \cdot n$ and, for any pair of morphisms m' of $\hom_A(\mathrm{D}', \mathrm{A})$, n' of $\hom_A(\mathrm{D}', \mathrm{B})$ with $p \cdot m' = q \cdot n'$, there exists a unique morphism t from D′ to D such that $m \cdot t = m'$ and $n \cdot t = n'$.

This situation can also be described differently in terms of a limit of a suitable functor from a small category. For the small category J we take one with three objects (1, 2, 3) and five morphisms (the three identity morphisms at 1, 2, 3 and an f from 1 to 2 and a g from 3 to 2). If now F is a functor from this J to A, it is not hard to see that F has a limit (B^*, p^*) in A iff $(p^*(1), p^*(3))$ is a pullback of $[F(f), F(g)]$. (See Figure 6.3.)

Figure 6.2

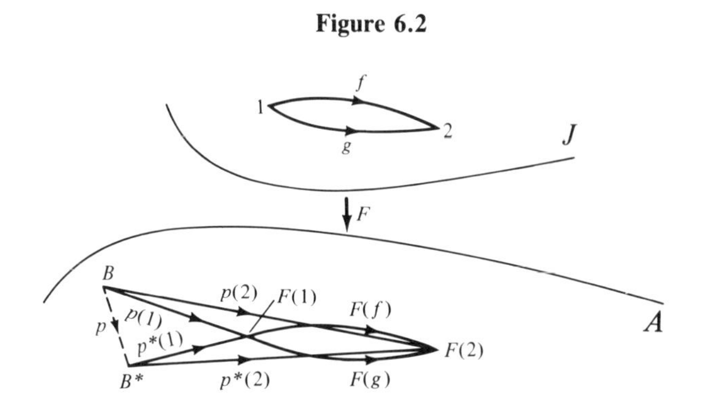

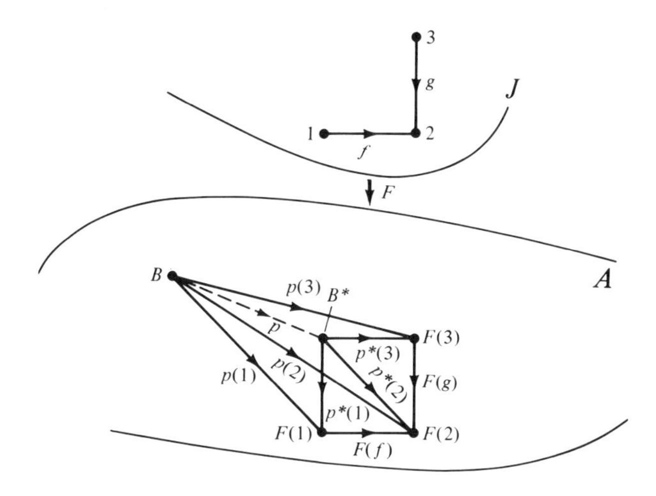

Figure 6.3

Before considering examples we relate the pullback to products and equalizers.

Lemma 6.1.1 *(i) For f, g from hom_A(A, B), (C, e) is an equalizer if (e, e) is a pullback of (f, g) with dom(e) = C.*

(ii) When the product A × B *exists and $p(1)$, $p(2)$ denote the canonical morphisms from* A × B *to* A, B, *respectively, for morphisms f in hom_A(A, C) and g in hom_A(B, C), ($p(2) \cdot e$, $p(1) \cdot e$) will be a pullback of (f, g) if* (E, e) *is an equalizer of ($f \cdot p(1)$, $g \cdot p(2)$).*

(iii) When (m, n) is a pullback of (p, q), the morphism m is a monomorphism, a regular monomorphism, or a retraction whenever q is such.

PROOF. (i) Assuming (e, e) is a pullback for (f, g), we have $f \cdot e = g \cdot e$; and when $f \cdot k = g \cdot k$, there is a unique h such that $e \cdot h = k$. This means that (dom(e), e) is an equalizer of (f, g).

(ii) If (E, e) is an equalizer of ($f \cdot p(1)$, $g \cdot p(2)$), then $f \cdot p(1) \cdot e = g \cdot p(2) \cdot e$; further, if $h(1)$ in hom_A(H, A) and $h(2)$ in hom_A(H, B) satisfy the relation $f \cdot h(1) = g \cdot h(2)$ (as in Figure 6.4), from the definition of (D, p) as a product of (A, B) there must be a unique h in hom_A(H, D) such that $p(1) \cdot h = h(1)$ and $p(2) \cdot h = h(2)$. This gives also $f \cdot p(1) \cdot h = g \cdot p(2) \cdot h$. Since (E, e) was taken as an equalizer for the pair ($f \cdot p(1)$, $g \cdot p(2)$), it follows that there is a unique k in hom_A(H, E) such that $e \cdot k = h$. We can see that such a k satisfies $p(1) \cdot e \cdot k = h(1)$ and $p(2) \cdot e \cdot k = h(2)$. Further, if a k' in hom_A(H, E) satisfies the relations $p(1) \cdot e \cdot k' = h(1)$ and $p(2) \cdot e \cdot k' = h(2)$, the uniqueness we asserted for h would imply that $e \cdot k' = h = e \cdot k$; and then the uniqueness we asserted for

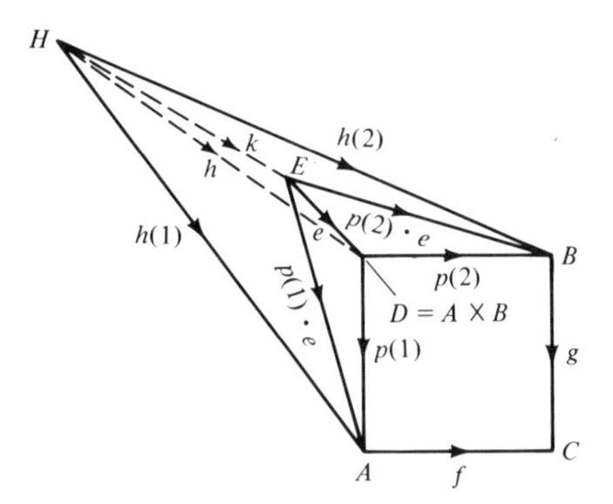

Figure 6.4

k would give $k' = k$. Thus we have shown that $(p(1) \cdot e\ , p(2) \cdot e)$ is indeed a pullback for (f, g).

(iii) Let (m, n) be a pullback of the pair (p, q): m: D → A, n: D → B, p: A → C, and q: B → C. (See Figure 6.5.)

Assume first that q is a monomorphism: if r, s from $\hom_A$(E, D) satisfy $m \cdot r = m \cdot s\ (=m')$, then $q \cdot (n \cdot r) = (q \cdot n) \cdot r = (p \cdot m) \cdot r = p \cdot (m \cdot r) = p \cdot (m \cdot s) = (p \cdot m) \cdot s = (q \cdot n) \cdot s = q \cdot (n \cdot s)$. Since q is left-cancellative, we get then $n \cdot r = n \cdot s\ (=n')$. Since m': E → A and n': E → B satisfy $p \cdot m' = q \cdot n'$, the assumption that (m, n) is a pullback for (p, q) implies that there is a unique h: E → D such that $m \cdot h = m'$ and $n \cdot h = n'$. Also, $m \cdot r = m'$, $n \cdot r = n'$ and $m \cdot s = m'$, $n \cdot s = n'$; hence, the uniqueness part of h gives $r = h = s$. This finally proves that m is left-cancellative or a monomorphism.

Next, let (B, q) be an equalizer of a pair of morphisms (r', s') with common domain C and a common codomain. Then $(r' \cdot p) \cdot m = r' \cdot (p \cdot m) = r' \cdot (q \cdot n) = (r' \cdot q) \cdot n = (s' \cdot q) \cdot n = s' \cdot (q \cdot n) =$

Figure 6.5

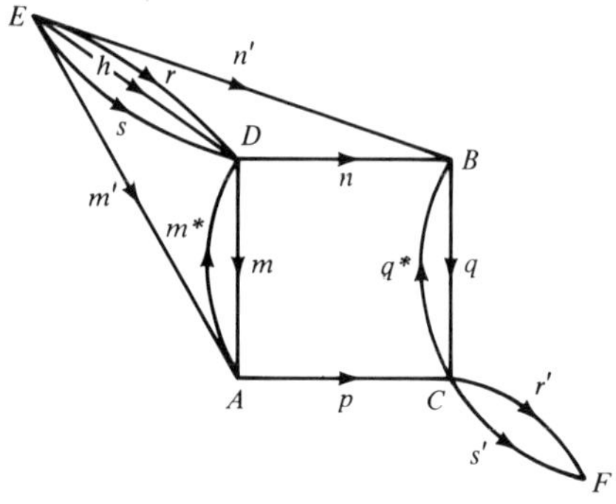

$s' \cdot (p \cdot m) = (s' \cdot p) \cdot m$. If for an m': E → A we have $(r' \cdot p) \cdot m' = (s' \cdot p) \cdot m'$, from the assumption that (B, q) is an equalizer for (r', s') would follow that there is a unique n': E → B such that $p \cdot m' = q \cdot n'$. Then, since (m, n) is a pullback for (p, q), it would follow that there is a unique h: E → D with $m \cdot h = m'$ and $n \cdot h = n'$. If, finally, for an h': E → D we have $m' = m \cdot h'$, $q \cdot (n \cdot h') = (q \cdot n) \cdot h' = (p \cdot m) \cdot h' = p \cdot m' = q \cdot n'$ would give, since q is a (regular) monomorphism, that $n \cdot h' = n'$ also. Hence, the uniqueness statement for h would give $h' = h$. Thus we have a unique h with $m \cdot h = m'$ whenever $(r' \cdot p) \cdot m' = (s' \cdot p) \cdot m'$. This shows that (D, m) is an equalizer of $(r' \cdot p, s' \cdot p)$. Thus when q is a regular monomorphism, so is m.

Finally, assume that q is a retraction, and therefore has a right inverse q^*: $q \cdot q^* = 1_C$. From $p \cdot 1_A = p = q \cdot (q^* \cdot p)$, and the assumption that (m, n) is an equalizer for (p, q), it follows that there is a unique m^*: A → D such that $m \cdot m^* = 1_A$ and $n \cdot m^* = q^* \cdot p$. It follows that m has a right inverse and so is a retraction. □

Part (ii) of the lemma gives the typical form of a pullback in many familiar categories. For instance, in a category *FA* of *F*-algebras and their *F*-homomorphisms, if f: A → C and g: B → C are given homomorphisms, a pullback for (f, g) would be $(p(1) \cdot e, p(2) \cdot e)$ where e: E → (A × B) is the inclusion morphism from the subalgebra E of (A × B) consisting of those (a, b) for which $f(a) = g(b)$. Similar remarks apply to the case of continuous maps f: A → C, g: B → C in the category *T* of topological spaces.

Part (iii) shows that when we have two subalgebras (or two subspaces), A, B of an algebra C (or space C), the pullback for their inclusion maps is a pair of inclusion maps from the intersection of the subalgebras or subspaces.

6.1.4. The Intersection of Subobject Families

A family of subobjects of an object A in a category *A* would correspond to a family of monomorphisms all with the same codomain A. So let us take a small category *J* in which all the nonidentity morphisms have a common codomain f; say they are of the form $m(j)$: $j \to f$, with an indexing set $[j]$, which is effectively the set of domains of these morphisms. We may call such a small category a *sink* (following Herrlich and Strecker [15] in their book on category theory: the image set of objects and morphisms of such a *J* by a functor in a category *A* is what they call a sink in *A*).

Let *F* now be a functor from such a sink category *J* to *A*, and let *F* map the morphisms $m(j)$ of *J* into monomorphisms of *A*, so that the $[F(j)]$ form a set of subobjects of $F(f)$. If this *F* has a limit (B*, p^*), we shall show that the $[p^*(j)]$ and $p^*(f)$ are all monomorphisms; $(B^*, p^*(f))$ is then defined as an *intersection* of the subobjects $[F(j), F(m(j))]$. It would be clearly unique up to an isomorphism in A. To prove $p^*(j)$, $p^*(f)$ are monomorphisms (see Figure 6.6): let r, s be from $\hom_A(E, B^*)$ with $p^*(f) \cdot r = p^*(f) \cdot s$. Since $F(m(j)) \cdot p^*(j) = p^*(f)$ for each j, we get $F(m(j)) \cdot p^*(j) \cdot r = F(m(j)) \cdot p^*(j) \cdot s$ for each j. Since the $F(m(j))$ are monomorphisms, by assumption, it follows that $p^*(j) \cdot r = p^*(j) \cdot s$ for each j. Setting $p'(f) = p^*(f) \cdot r = p^*(f) \cdot s$ and $p'(j) = p^*(j) \cdot r = p^*(j) \cdot s$, we see that (E, p') is a predecessor of F; hence, from the definition of the limit (B^*, p^*) it follows that there is a unique $h: E \to B^*$ such that $p'(f) = p^*(f) \cdot h$ and $p'(j) = p^*(j) \cdot h$ for each j. But both r and s satisfy the conditions here for h, so its uniqueness implies that $r = s$. It follows that $p^*(f)$ is a monomorphism. From $p^*(f) = F(m(j)) \cdot p^*(j)$ it then follows that $p^*(j)$ is also a monomorphism.

In familiar examples like T, or S or FA, or FDA, an intersection of a family of subobjects of an object has a set base that is the intersection of the set bases of the subobjects.

Figure 6.6

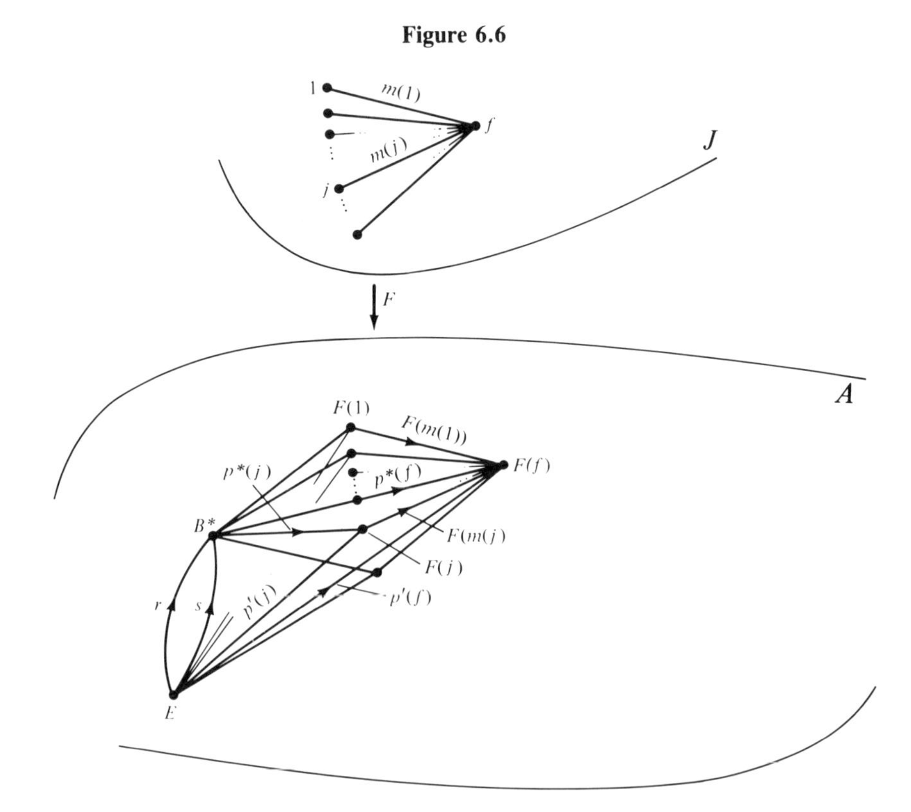

6.1.5. Limits for Inverse-Directed Families

An ordered set $(J, \leq)$ is said to be inversely directed when it is down-directed (that is, any two elements have a common lower bound). Such an inverse-directed (ordered) set $(J, \leq)$ can be viewed, in a natural manner, as a small category (with J as set of objects) in which [$\hom_J(j, k)$ is nonnull and contains exactly one element] iff [$j \leq k$].

When we take such an "inverse-directed set category" J and a functor F from J to a category A, the image objects and morphisms in A (under F) of the objects and morphisms in J give rise to an *inverse-directed family* in A. If $F: J \to A$ has a limit (B^*, p^*) in A, we call it a *limit of the inverse-directed family* $F(J)$. We shall meet an important type of such a limit in a later section.

6.2 SUCCESSORS AND COLIMITS OF A FUNCTOR

These dual notions are defined in the obvious fashion. Given a functor $F: J \to A$ where J is assumed to be a small category, a successor (q, C) of F is a pair consisting of an object C of A and a natural transformation $q: F \to c^*(C)$ from F to the constant functor $c^*(C): J \to A$. These successors form the objects of a category $Su(F)$, with $\hom_{Su(F)}[(q, C), (q', C')]$ consisting of all morphisms $m^*: C \to C'$ such that $m^* \circ q(j) = q'(j)$, for each j of $O(J)$. If this category $Su(F)$ should have an initial object (q^*, C^*), it is called a *colimit of F in A*. Such a colimit would be unique up to isomorphisms in $Su(F)$, hence, effectively, up to isomorphs of C^*.

Then, corresponding to the various examples of such limits in the last section, we have dual notions for colimits. Briefly, they are the following:

The Coproduct. The coproduct is a colimit of a functor F in A from a small discrete category J. Familiar examples are the disjoint union in S and the disjoint sum in T.

The Coequalizer. The coequalizer is the colimit of a functor F in A from a small category J with two objects (1, 2), and two nonidentity morphisms f, g in $\hom_J(1, 2)$. It would be an epimorphism from $F(2)$ called a regular epimorphism. Duals of earlier results proved for equalizers are true for coequalizers.

The Push-out. The push-out is defined as a colimit of a functor F in A, from a small category J, with three objects (1, 2, 3) and two nonidentity morphisms $f: 2 \to 1$, $g: 2 \to 3$. Duals of results stated in Lemma 6.1.1 can be proved now for this. Usually we say that the pair of morphisms $q^*(1)$, $q^*(3)$ form a push-out for the pair of morphisms $F(f)$, $F(g)$ if (q^*, B^*) is a colimit of F in A.

The Cointersection. Let J be a "source category" having a set of morphisms, all the nonidentities having the same domain d. If a functor $F: J \to A$ associates to these nonidentities epimorphisms in A as images, then a colimit (q^*, B*) of such an F in A is called a cointersection of the family of "quotient objects" of $F(d)$ given by the family of the image epimorphisms. For example, in a category like FA if D is the F-algebra $F(d)$ and the quotient algebras are given by the congruences $[E(j) : j$ in $J_0]$, cointersection would be given by taking the quotient by the congruence generated by the union of the $[E(j)]$.

The Colimit of a Directed Family. When $(J, \leq)$ is a directed (ordered) set (that is, an ordered set in which each pair of elements has a common upper bound), then considering this as a small category J, if a functor F: $J \to A$ has a colimit (q^*, B*), we call B* a *colimit of the directed family of objects* $[F(j)$: j in $J]$. In earlier literature this colimit is often called a direct limit, in contrast to the dually defined inverse limit.

6.3 FACTORIZATIONS OF MORPHISMS

In the category S of sets and set maps, when f: A $\to$ B is a given set map, we have a subset f(A) $=$ [b in B: $b = f(a)$ for some a of A] of B and an equivalence relation $E = [(a, a')$ in A $\times$ A: $f(a) = f(a')]$ on A. If I denotes the inclusion map of f(A) in B, there is a surjective map f_1: A $\to f$(A), with $f_1(a) = f(a)$ for each a of A, such that $f = I \cdot f_1$; also, if θ is the canonical surjection of A on A/E, there is an injective map m: A/$E \to$ B such that $f = m \cdot \theta$. Each of these gives a factorization of the general morphism f of S as a composition of an epimorphism followed by a monomorphism. We also have a certain "uniqueness" for such a factorization. If $f = m \cdot e$ and $m' \cdot e'$ with m, m' as monomorphisms and e, e' as epimorphisms, then there is an isomorphism (or bijective map) i such that $i \cdot e = e'$ and $m' \cdot i = m$. We seek general results regarding the possibility of such factorizations and their possible uniqueness.

We begin by identifying a class of epimorphisms weaker than the regular ones, called *extremal epimorphisms* and dually *extremal monomorphisms*. We consider, too, a "diagonalizing" property that helps to establish uniqueness of factorizations.

Definition 6.3.1 An epimorphism e in the category C is called an *extremal epimorphism* if [$e = m \cdot g$ with m a monomorphism] implies [m is an isomorphism]; and dually, we define an extremal monomorphism.

Definition 6.3.2 A brace of pairs [m, e; m', e'] of morphisms from the category C is called a "square" of morphisms if $m \cdot e = m' \cdot e'$. Such a square is said to be diagonalizable if there is a (diagonal) morphism h such that $h \cdot e = e'$ and $m' \cdot h = m$.

Lemma 6.3.1 *(a) Any regular epimorphism is an extremal epimorphism.*

(b) The category C is a balanced category iff either (i) each epimorphism is an extremal epimorphism, or (ii) each monomorphism is an extremal monomorphism.

(c) A square $[g, e; m, f]$ is diagonalizable if either (i) e is a regular epimorphism and m a monomorphism, or (ii) m is a regular monomorphism and e is an epimorphism.

PROOF. (a) Given e is a regular epimorphism, let it be the coequalizer of a pair of morphisms (r, s) in $\hom_C(A, B)$; we assume e is in $\hom_C(B, C)$. To show that e is an extremal epimorphism, suppose that $e = m \cdot h$ for m a monomorphism of C; we have to prove that m is an isomorphism. Because e is a coequalizer of r, s, we have $m \cdot h \cdot r = e \cdot r = e \cdot s = m \cdot h \cdot s$. Since m as a monomorphism is left-cancelable, we get then $h \cdot r = h \cdot s$. Hence, from the property of a coequalizer there is a morphism g such that $g \cdot e = h$. Thus $1_C \cdot e = e = m \cdot h = m \cdot g \cdot e$; and canceling the epimorphism e from the right, we get $1_C = m \cdot g$. Hence m has a right inverse, so it is a retraction as well as a monomorphism. But then it must be an isomorphism.

(b) If C is a balanced category and e is an epimorphism in C with a factorization $e = m \cdot g$ where m is a monomorphism, $r \cdot m = s \cdot m$ would imply $r \cdot e = r \cdot m \cdot g = s \cdot m \cdot g = s \cdot e$, and so $r = s$. Thus m is also an epimorphism. Since C is balanced, m must be an isomorphism. That is, e is extremal. Conversely, if each epimorphism in C is extremal, and e is an epimorphism as well as a monomorphism, [$e = e \cdot 1$, where the first e is an extremal epimorphism and the second e is a monomorphism] would give [the second e is an isomorphism]. Thus C is balanced. The dual is equally true.

(c) Suppose that e is a regular epimorphism and m a monomorphism in the square of morphisms $[g, e; m, f]$. That is, e is a coequalizer of a pair of morphisms (r, s). Then $e \cdot r = e \cdot s$ gives $m \cdot f \cdot r = g \cdot e \cdot r = g \cdot e \cdot s = m \cdot f \cdot s$ (since $g \cdot e = m \cdot f$). Since m is left-cancelable, this gives $f \cdot r = f \cdot s$. The definition of a coequalizer implies then that $f = h \cdot e$ for a morphism h. Then $m \cdot h \cdot e = m \cdot f = g \cdot e$. Because e is right-cancelable, we get $m \cdot h = g$. This proves that there is a diagonal morphism h. The dual can be similarly proved. □

In the category C we suppose that E is a class of epimorphisms of C closed for composition with isomorphisms (on either side) and M a class of monomorphisms also closed for composition with isomorphisms.

Definition 6.3.3 We say that C *admits (E, M)-factorizations* if each f in $M(C)$ can be expressed in the form $f = m \cdot e$ for m, e from M, E. We say that C *admits unique (E, M)-factorizations* if it admits (E, M)-factorizations and when $f = m \cdot e = m' \cdot e'$ for m, m' from M and e, e' from E, there is an isomorphism i such that $i \cdot e = e'$ and $m' \cdot i = m$; and we call C an *(E, M)-category* if C admits unique (E, M)-factorizations and further each of M, E is closed for composition.

Theorem 6.3.1 *(a) If C admits (E, M)-factorizations, C will be an (E, M)-category iff each square of morphisms $[g, e; m, f]$ with m, e from M, E is diagonalizable.*

(b) If C is well-powered, every set of subobjects of an object has an intersection, and every pair of parallel morphisms (with common domain and codomain) has an equalizer, then C admits (E, M)-factorizations when E is the class of extremal epimorphisms and M the class of monomorphisms.

(c) If C admits (E, M)-factorizations with E, M as in (b), then it admits unique (E, M)-factorizations provided that each pair (f, g) of $O(C)$ with a common codomain has a pullback.

PROOF. (a) Assuming first that C is an (E, M)-category, we prove that a square of morphisms $[g, e; m, f]$ with e in E and m in M is diagonalizable: let $f = m' \cdot e'$ and $g = m'' \cdot e''$ be (E, M)-factorizations for f and g. Since $g \cdot e = m \cdot f$ we get $m'' \cdot e'' \cdot e = m \cdot m' \cdot e'$; since $M \cdot M \subseteq M$ and $E \cdot E \subseteq E$, $m \cdot m'$ is in M and $e'' \cdot e$ is in E. Thus from $m'' \cdot (e'' \cdot e) = (m \cdot m') \cdot e'$ we have an isomorphism i in C such that $i \cdot (e'' \cdot e) = e'$ and $(m \cdot m') \cdot i = m''$. Setting $h = m'\, i \cdot e''$, we see then that $h \cdot e = m' \cdot e' = f$ and $m \cdot h = m'' \cdot e'' = g$. Thus the square $[g, e; m, f]$ has been shown to be diagonalizable. (See Figure 6.7.)

Conversely, assuming that C admits (E, M)-factorizations and that each square $[g, e; m, f]$ with e in E, and m in M is diagonalizable, we now show that C is an (E, M)-category. First, if we have two equal factori-

Figure 6.7

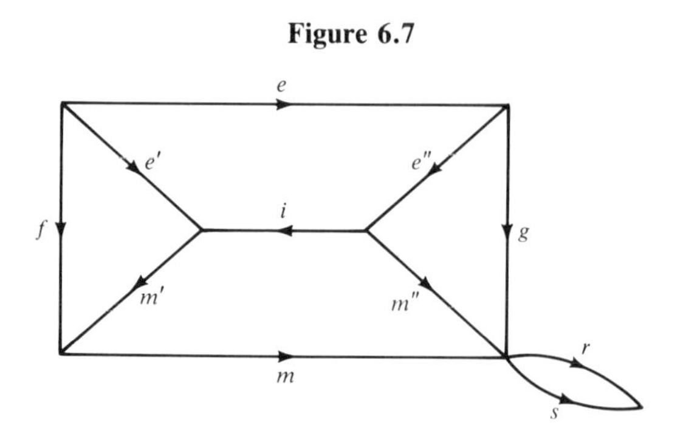

zations $f = m \cdot e = m' \cdot e'$ of the same f, with m, m' from M and e, e' from E, then the diagonalization property applies to $[m, e; m', e']$ and to $[m', e'; m, e]$, so that there are morphisms h and h' such that $h \cdot e = e'$, $m' \cdot h = m$, and $h' \cdot e' = e$, $m \cdot h' = m'$. It follows that $h' \cdot h \cdot e = e = 1_C \cdot e$ and $m' \cdot h \cdot h' = m' = m' \cdot 1_D$, so that, e being right-cancelable and m' being left-cancelable, we get $h' \cdot h = 1_C$, $h \cdot h' = 1_D$. Thus h and h' are isomorphisms. This proves the essential uniqueness of an (E, M)-factorization of a morphism of C. Next we check that M and E are both closed for composition of morphisms in them. We take the case of M; the other may be proved dually. If m_1, m_2 are from M and have a composition $m_1 \cdot m_2$, resolving this $m_1 \cdot m_2$ as $m \cdot e$, we obtain from the diagonalizing of $[m, e; m_1, m_2]$ a morphism h such that $h \cdot e = m_2$ and $m_1 \cdot h = m$. From the first relation, the square $[h, e; m_2, 1_A]$ has a diagonalization. Thus there is an h' such that $h' \cdot e = 1_A$, $m_2 \cdot h' = h$. Since e has a left-inverse, it is a section. Because e is also an epimorphism, it must be an isomorphism. But then $m \cdot e$, as a composite of an element of m with an isomorphism, belongs to M. That is, $m_1 \cdot m_2 = m \cdot e$ belongs to M.

(b) Using the given hypotheses, we shall get an (E, M)-factorization for a given f: A $\to$ B in C. First, we select a representative set of subobjects $[\mathrm{B}(k), m(k)$: k in $K]$ of B, assuming that this set includes $(\mathrm{B}, 1_\mathrm{B})$; say $(\mathrm{B}, 1_\mathrm{B}) = (\mathrm{B}(o), m(o))$ for an o in K. From the set K we next choose those elements j for which $m(j)$ is a left-factor of f, so that $f = m(j) \cdot g(j)$ for some $g(j)$ of $M(C)$. The element o of K is such a j, since $f = m(o) \cdot f$. Let J denote the subset of these j's from K. Our hypothesis then provides an intersection for the set of subobjects $[\mathrm{B}(j)$: j in $J]$. That means that if J^* denotes the small subcategory of C with the objects $[\mathrm{B}(j)]$ and the morphisms [the identities $1_{\mathrm{B}(J)}$ and the morphisms $m(j)$], and I denotes the inclusion functor from J^* to C, then this I has a limit (B^*, p^*) in C. (See Figure 6.8.) Then $(\mathrm{B}^*, p^*(\mathrm{B}(o)))$ will be an intersection of the set of subobjects $[\mathrm{B}(j), m(j)]$. Hence $p^*(\mathrm{B}(O)) = m(j) \cdot p^*(\mathrm{B}(j))$ for each j in J. We now have another predecessor (A, g^*) of I in C when we set $g^*(\mathrm{B}(o)) = f$ and $g^*(\mathrm{B}(j)) = g(j)$ for each j of J, where f has an assigned resolution $f = m(j) \cdot g(j)$, by our choice of J. (In fact, for $j = o$, this $g(j)$ is f, too.) The fact that $f = m(j) \cdot g(j)$ for each j makes g^* a natural transformation from $c^*(\mathrm{A})$ to I, and so (A, g^*) a predecessor of I. From the property of the limit of I in C there must be a unique morphism e from A to B* such that $p^*(\mathrm{B}(j)) \cdot e = g(j)$ for each j in J. In particular, $f = p^*(\mathrm{B}(o)) \cdot e$ (for $j = o$). Note that $p^*(\mathrm{B}(o))$ is a monomorphism. We prove that e is an epimorphism and an extremal one. To prove e is an epimorphism, let $r \cdot e = s \cdot e$ for r, s from $\hom_B(\mathrm{B}, \mathrm{D})$. (See Figure 6.8.) This pair (r, s) has an equalizer (C, m) by our hypothesis, and m is a monomorphism too. The property of the equalizer implies (since $r \cdot e = s \cdot e$) that there is a unique h in $\hom_B(\mathrm{A}, \mathrm{C})$ such that $m \cdot h = e$. From $f = p^*(\mathrm{B}(o)) \cdot e =$

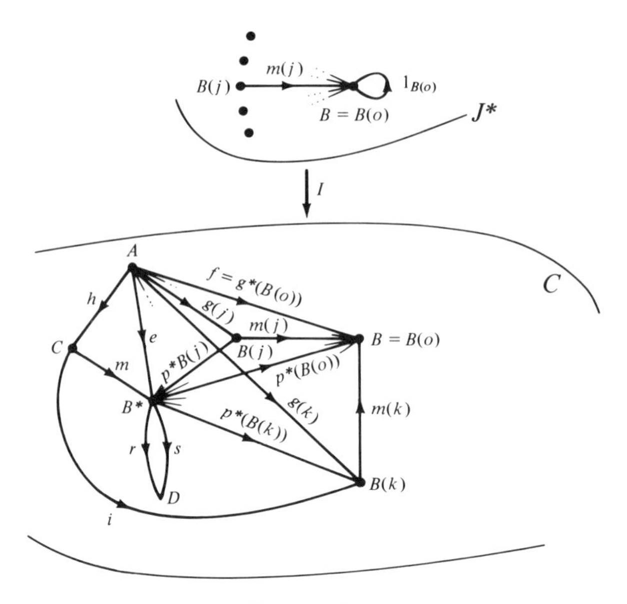

Figure 6.8

$p^*(B(o)) \cdot m \cdot h$ since $p^*(B(o)) \cdot m$ is a monomorphism, it follows that $(C, p^*(B(o)) \cdot m)$ is isomorphic with one of the $(B(k), m(k))$, k in K; thus there is an isomorphism i: $C \to B(k)$ with $m(k) \cdot i = p^*(B(o)) \cdot m$ and $i \cdot h = g(k)$. From $p^*(B(o)) = m(k) \cdot p^*(B(k))$ we get $p^*(B(o)) \cdot 1_{B^*} = m(k) \cdot i \cdot i^r \cdot p^*(B(k)) = p^*(B(o)) \cdot m \cdot i^r \cdot p^*(B(k))$. Because $p^*(B(o))$ is a monomorphism, it is left-cancelable, so we get $1_{B^*} = m \cdot i^r \cdot p^*(B(k))$. Hence m has a right inverse and is a retraction as well as a monomorphism. Thus it is an isomorphism. But then, since (C, m) is an equalizer for (r, s), this means that $r = s$. Thus e is an epimorphism. To see that it is an extremal epimorphism, let $e = m \cdot h$ for some monomorphism m (see Figure 6.8). Again $(C, p^*B(o)) \cdot m)$ is isomorphic with a $(B(k), m(k))$, and we have the same i as before. Hence $i \cdot h = g(k) = p^*(B(k)) \cdot e$ and $1_{B^*} \cdot e = e = m \cdot h = m \cdot i^r \cdot i \cdot h = m \cdot i^r \cdot p^*(B(k)) \cdot e$; but we know that e is an epimorphism, and therefore right-cancelable. Thus we get $1_{B^*} = m \cdot (\cdots)$, showing that m is a retraction as well as a monomorphism. Hence it is an isomorphism, proving that e is an extremal epimorphism. These assertions prove (b).

(c) Now assuming that C admits (E, M)-factorizations with E consisting of extremal epimorphisms and M of monomorphisms, and further that pullbacks exist for each pair of morphisms in C with common domain, we prove that a square of morphisms $[g, e; m, f]$, where e is an extremal

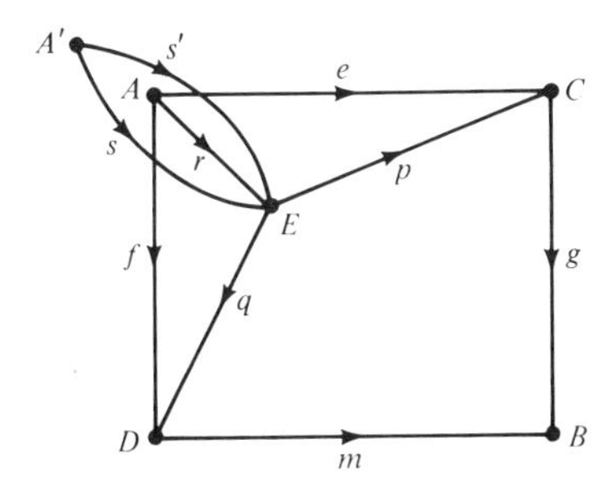

Figure 6.9

epimorphism and m is a monomorphism, is diagonalizable. Then by part (a) it would follow that C admits unique (E, M)-factorizations.

Let (p, q) be a pullback of the pair (g, m), where g, m have a common codomain B. (See Figure 6.9.) Since $g \cdot e = m \cdot f$, the pullback definition implies that there is an r [in $\hom_C(A, E)$, as in the figure] such that $e = p \cdot r$ and $f = q \cdot r$. We claim that p must be a monomorphism: for if $p \cdot s = p \cdot s'$ for some s, s' from $\hom_C(A', E)$, $m \cdot q \cdot s = g \cdot p \cdot s = g \cdot p \cdot s' = m \cdot q \cdot s'$; since m is left-cancellative, we have then $q \cdot s = q \cdot s'$ besides $p \cdot s = p \cdot s'$. From the characteristic property of the pullback, denoting $p \cdot s = p \cdot s'$ as p' and $q \cdot s = q \cdot s'$ as q', the fact that $g \cdot p' = m \cdot q'$ implies that there is a unique k such that $p \cdot k = p'$ and $q \cdot k = q'$. Hence $s = k = s'$. Thus $s = s'$ follows from $p \cdot s = p \cdot s'$. Thus p is a monomorphism. But then from $e = p \cdot r$ and e is an extremal epimorphism it follows that p must be an isomorphism. We now set $h = q \cdot p^{r}$; then $h \cdot e = q \cdot p^{r} \cdot e = q \cdot p^{r} \cdot p \cdot r$ (since $e = p \cdot r$) $= q \cdot r = f$; and so $m \cdot h \cdot e = m \cdot f = g \cdot e$. Since e is right-cancellative this last gives $m \cdot h = g$ also. Hence we have the required diagonal morphism h. □

A dual result provides (epimorphism, extremal monomorphism)-factorizations in a suitable category C. But even with the hypothesis in part (b), Herrlich and Strecker ([15], Chapter 9, Theorem 34.5) have proved the same C also admits unique (E^*, M^*)-factorizations where E^*, M^* are the classes of epimorphisms and extremal monomorphisms in C.

6.4 COMPLETENESS

We have been looking at various types of limits of a functor, from various types of small categories. Now we examine some interrelations among them and look for some all-inclusive types.

For a given small category J, if every functor from J to C has a limit in C, we say that *C has J-limits*. By suitably restricting J we get special cases of this. Thus, we say that

1. *C has finite products* if it has J-limits for each finite discrete category J;
2. *C has equalizers* if C has J-limits for J equal to the small category with two objects (1, 2) and two nonidentity morphisms both from 1 to 2;
3. *C has pullbacks* if C has J-limits for J equal to the small category with three objects (1, 2, 3) and two nonidentity morphisms f: $1 \rightarrow 2$, g: $3 \rightarrow 2$;
4. *C has products* if C has J-limits for each small discrete category J;
5. *C has inverse limits* if C has J-limits whenever J is an inverse-directed (ordered) set considered as a category, in the usual way;
6. *C is finitely complete* if C has J-limits for each finite category J;
7. *C is complete* if C has J-limits for each small category J.

(Dually we define cocomplete and finitely cocomplete categories.)

Clearly the last property of completeness is the most inclusive. We begin by interrelating some of the others.

Lemma 6.4.1 *Of the following three conditions on a category C, we have the relations (a) $\Leftrightarrow$ (b) $\Rightarrow$ (c):*

(a) C has finite/arbitrary products and pullbacks;
(b) C has finite/arbitrary products and equalizers;
(c) any finite/arbitrary set of regular subobjects (B(k), m(k)) *of an object* A *of C has an intersection* (B, m) *that is also a regular subobject of* A. [*We call* (B, m) *a regular subobject of* A *if m:* B $\rightarrow$ A *is a regular monomorphism*].

PROOF. In part (ii) of Lemma 6.1.1 we proved that (b) implies (a). To prove that (a) implies (b), we first note that when (A $\times$ B, p) denotes a limit of the functor F from the discrete category (1, 2) in C with $F(1) =$ A, $F(2) =$ B (so that A $\times$ B is what is called a direct product of A and B), then whenever there is an object D of C with morphisms d_1: D $\rightarrow$ A and d_2: D $\rightarrow$ B in C, (D, d) may be treated as a predecessor of F [with $d(i) = d_i$, $i = 1$ or 2], so there is a unique morphism d^*: D $\rightarrow$ A $\times$ B such that $p(i) \cdot d^* = d_i$ ($i = 1, 2$). Assuming now that we are given a functor F from J to C, where J has two objects (1, 2) and a pair of nonidentity morphisms (f, g) both from 1 to 2, we suppose that there is product ($F(1) \times F(2)$, p) for the pair of objects $F(1)$, $F(2)$. Suppose further that the pair of morphisms $(1_{F(1)}, F(f))$ and $(1_{F(1)}, F(g))$ both from $F(1)$ to $F(1) \times F(2)$ gives rise to a pullback q_1: C $\rightarrow F(1)$, q_2: C $\rightarrow F(2)$. (See Figure 6.10.) Then $p(1) \cdot (1_{F(1)}, F(f)) \cdot q_1 = p(1) \cdot (1_{F(1)}, F(g)) \cdot q_2$ and $p(2) \cdot (1_{F(1)}, F(f)) \cdot q_1 = p(2) \cdot (1_{F(1)}, F(g)) \cdot q_2$ give us $1_{F(1)} \cdot q_1 = 1_{F(1)} \cdot q_2$ and

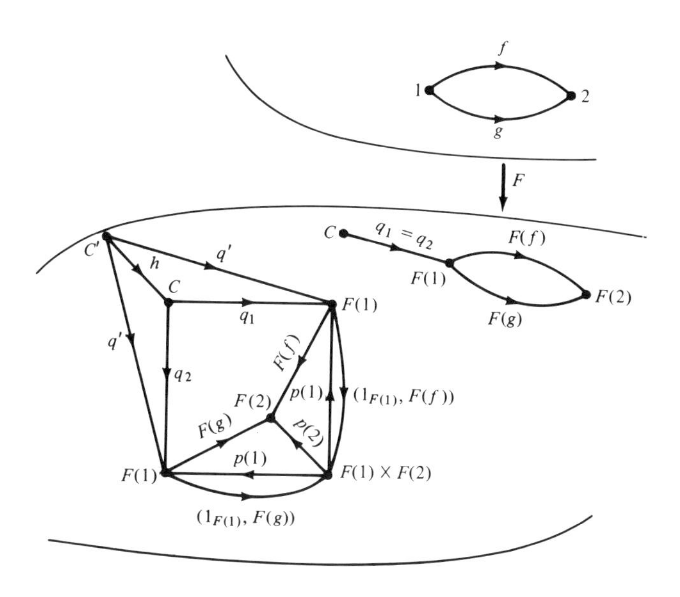

Figure 6.10

$F(f) \cdot q_1 = F(g) \cdot q_2$. Hence $q_1 = q_2$ is a morphism from C to $F(1)$ such that $F(f) \cdot q_1 = F(g) \cdot q_1$; further, if $F(f) \cdot q' = F(g) \cdot q'$ for a morphism q' from a C′ to $F(1)$, then $(1_{F(1)}, F(f)) \cdot q' = (1_{F(1)}, F(g)) \cdot q'$ would imply, from the pullback assumption, that there is a unique morphism h from C′ to C such that $q_1 h = q'$. Thus (C, q_1) is an equalizer of the pair $F(f)$, $F(g)$ (or a limit of F in C).

Finally we prove that (b) implies (c). Let us assume that C has equalizers and has finite/arbitrary products. (See Figure 6.11.) If [(D_i, e_i)] is a finite/arbitrary family of regular subobjects of an object D_o in C, with (D_i, e_i) an equalizer of a pair (r_i, s_i) of morphisms from D_o to A_i for each i, let (A, p) be a product of the family (A_i). The family of morphisms r_i: $\mathrm{D}_o \to \mathrm{A}_i$ then determine a unique semiproduct morphism r: $\mathrm{D}_o \to \mathrm{A}$ such that $p_i \cdot r = r_i$ for each i. Similarly, we have a semiproduct s: $\mathrm{D}_o \to \mathrm{A}$ of the s_i with $p_i \cdot s = s_i$ for each i. If (D, e) be an equalizer for this pair (r, s), both from D_o to A, we shall prove that (D, e) is an intersection of the family (D_i, e_i): for from the definition of (D, e) we have $r \cdot e = s \cdot e$, and so $r_i \cdot e = p_i \cdot r \cdot e = p_i \cdot s \cdot e = s_i \cdot e$ for each i. Since (D_i, e_i) is an equalizer for (r_i, s_i), there must be a unique morphism d_i: $\mathrm{D} \to \mathrm{D}_i$ such that $e_i \cdot d_i = e$. For the subcategory J of C whose objects are [D_o and the D_i] and whose morphisms are [the identities at D_o and the D_i, and the e_i] and the inclusion functor I of J in C we have a predecessor (D, d^*) if we set $d^*(\mathrm{D}_o) = e$ and $d^*(\mathrm{D}_i) = d_i$ for each i. To see that this

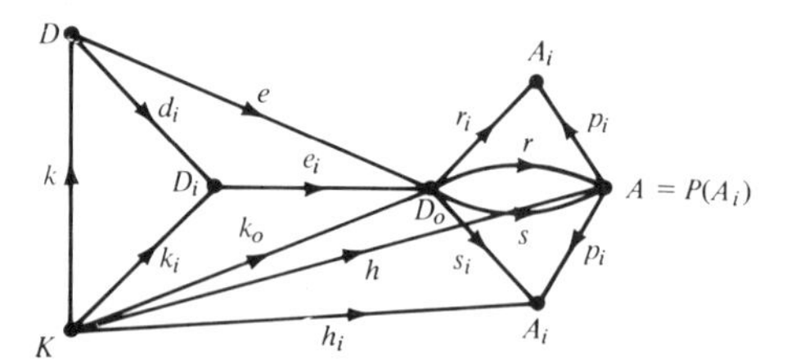

Figure 6.11

gives a limit for the functor I, suppose that (K, k^*) is any other predecessor of I; writing k_i for $k^*(\mathrm{D}_i)$ (for all i and for o too), we have $e_i \cdot k_i = k_o$ for the i. And $r_i \cdot k_o = r_i \cdot e_i \cdot k_i = s_i \cdot e_i \cdot k_i = s_i \cdot k_o$ for each i. Hence $p_i \cdot r \cdot k_o = p_i \cdot s \cdot k_o$ for each p_i. If we denote this common value by h_i, h_i: $\mathrm{K} \to \mathrm{A}_i$ are morphisms for which there is a semiproduct h, a unique h: $\mathrm{K} \to \mathrm{A}$, such that $p_i \cdot h = h_i$ for each i. Then $r \cdot k_o = h = s \cdot k_o$. Since (D, e) is an equalizer of (r, s), there must be a unique k: $\mathrm{K} \to \mathrm{D}$ such that $e \cdot k = k_o$, from which we get $e_i \cdot d_i \cdot k = e \cdot k = k_o = e_i \cdot k_i$ for each i. But e_i is left-cancelable; so $d_i \cdot k = k_i$ for each i follows.

These relations prove in effect that (D, e) is a limit of the functor I or is an intersection of the family of regular subobjects (D_i, e_i), and the intersection is also a regular subobject of D_o. □

With this lemma, we are now ready to work out conditions for a category to be finitely complete or complete.

Theorem 6.4.1 *(a) The following conditions are necessary and sufficient for the category C to be finitely complete:*

(1) C has J-limits for every finite category J;
(2) C has finite products and pullbacks;
(3) C has finite products and equalizers.

(b) The following conditions are necessary and sufficient for C to be complete;

(1′) C has J-limits for each small category J;
(2′) C has arbitrary products and pullbacks;
(3′) C has arbitrary products and equalizers;
(4′) C has finite products, equalizers, and inverse limits.

PROOF. In (a), condition (1) is just the definition of finite completeness. Surely (1) implies (2); Lemma 6.4.1 proves (2) and (3) are equivalent. We shall be proving that (3) implies (1), and with it also (3′) implies (1′); in (b), (1′) is the definition of completeness. Again it is obvious that (1′) implies (2′), and the equivalence of (2′) and (3′) is already in Lemma 6.4.1.

So we start with proving that (3)/(3′) implies (1)/(1′) (see Figure 6.12). Let C have finite/arbitrary products and equalizers. Let J be any finite/small category and F a functor from J to C. We denote by J^* the discrete subcategory of J having the same class of objects and by F^* the functor $F \cdot I$ from J^* to C where I is the inclusion functor from J^* to J. Our assumptions then give a limit (C, p^*) for this functor F^*. For a typical morphism $\theta: i \rightarrow j$ in J we have a pair of morphisms $F(\theta) \cdot p^*(i)$ and $p^*(j)$ from C to $F^*(j) = F(j)$. These have then an equalizer $(\mathrm{E}(\theta), d(\theta))$; $d(\theta)$: $\mathrm{E}(\theta) \rightarrow \mathrm{C}$, $F(\theta) \cdot p^*(i) \cdot d(\theta) = p^*(j) \cdot d(\theta)$. Then, using the last part of our Lemma 6.4.1, we have an intersection (E, d) for the family of regular subobjects $[(\mathrm{E}(\theta), d(\theta)$: θ in $M(J)]$. Thus there are morphisms $q(\theta)$: $\mathrm{E} \rightarrow \mathrm{E}(\theta)$ such that $d(\theta) \cdot q(\theta) = d$ for each θ. Setting $d^*(i) = p^*(i) \cdot d$ for each i in $J^* = O(J)$, we now show that (E, d^*) is a limit for the functor F in C. For the typical $\theta: i \rightarrow j$ in J we have $F(\theta) \cdot d^*(i) = F(\theta) \cdot p^*(i) \cdot d = F(\theta) \cdot p^*(i) \cdot d(\theta) \cdot q(\theta) = p^*(j) \cdot d(\theta) \cdot q(\theta) = p^*(j) \cdot d = d^*(j)$. Hence (E, d^*) is surely a predecessor of F. If next we have some predecessor (E', d') of F, so that for the typical $\theta: i \rightarrow j$ in J we have $F(\theta) \cdot d'(i) = d'(j)$, the family of morphisms $d'(i)$: $\mathrm{E}' \rightarrow F(i) = F^*(i)$ leads to a semi-product morphism d': $\mathrm{E}' \rightarrow \mathrm{C} = \mathrm{P}(F^*(i))$, so that $p^*(i) \cdot d' = d'(i)$ for each i. Then also $F(\theta) \cdot p^*(i) \cdot d' = F(\theta) \cdot d'(i) = d'(j) = p^*(j) \cdot d'$. Hence the choice of $(\mathrm{E}(\theta), d(\theta))$ as an equalizer implies that there is a (unique) morphism $d''(\theta)$: $\mathrm{E}' \rightarrow \mathrm{E}(\theta)$ such that $d(\theta) \cdot d''(\theta) = d'$ for each

Figure 6.12

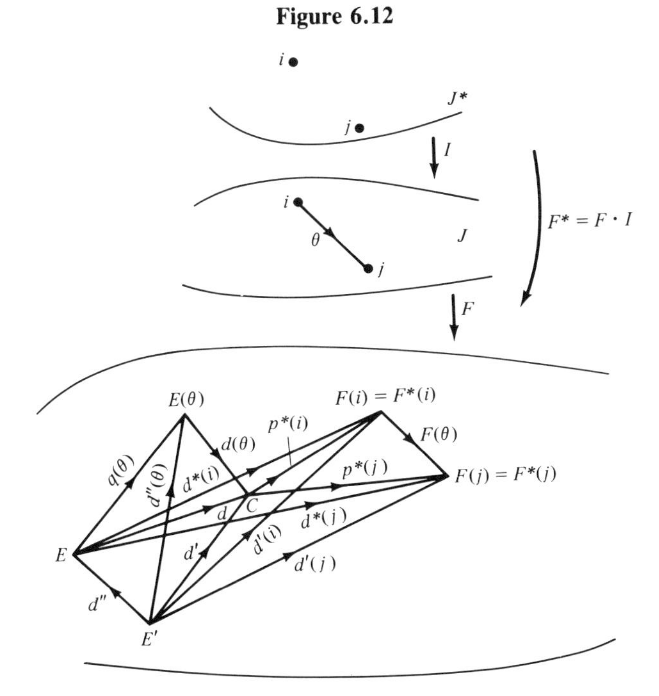

θ. (See Figure 6.12.) Now the definition of the intersection (E, d) of the (E(θ), $d(\theta)$) implies that there should be a morphism d'' from E$'$ to E such that $q(\theta) \cdot d'' = d''(\theta)$ for each θ. Hence we have $d^*(i) \cdot d'' = p^*(i) \cdot d \cdot d'' = p^*(i) \cdot d(\theta) \cdot q(\theta) \cdot d'' = p^*(i) \cdot d(\theta) \cdot d''(\theta) = p^*(i) \cdot d' = d'(i)$ for each i. This proves that (E, d^*) is indeed a limit for F.

Condition (1$'$) implies (4$'$) since inverse limits can be considered as cases of J-limits for special types of J. We prove the converse by checking that (4$'$) implies (3$'$). In fact, when C has finite products and inverse limits it has arbitrary products. This is what we now prove. Let $F: J \to C$ be a functor from the small discrete category J, where J may be considered as just an indexing set. To prove that this F has a limit in C we form an inverse-directed family ($FP(J)$, $\supseteq$) whose elements are the nonnull finite subsets of J; this is considered a (preordered) category with a unique morphism (A $\to$ B) whenever A $\supseteq$ B (and only then). We use the symbol (A $\to$ B) for this morphism. A functor F^* of this category in C can be defined by first choosing for each A of $FP(J)$ a product object $P[F(j)]$ of the $[F(j){:}j$ in A] in C with chosen canonical morphisms $p_A(j)$: $P[F(j)] \to F(j)$ for each j of A; then we set $F^*(A) = P[(F(j)]$ and $F^*(A \to B)$ equal to the semiproduct $p(A \to B)$ of the morphisms $[p_A(j){:}\ j$ in $B]$, that is, the unique $p(A \to B)$ for which $p_B(j) \cdot p(A \to B) = p_A(j)$ for each j of B. It is not hard to show that this F^* is a functor from the inverse-directed category ($PF(J)$, $\supseteq$) to C; our hypotheses then assure that this F^* has a limit (D, q) in C. If we set, for a j of J, $s(j) = q[(j)]$: D $\to P[F(j)] = F(j)$ [assuming that for a (j) of $FP(J)$, we had chosen $P[F(j)] = F(j)$ itself], we have a predecessor (D, s) of F in C. We claim that this is a limit of F. If (D$'$, k) were another predecessor of F, we would have a predecessor (D$'$, k') of F^* by setting $k'(A)$ equal to the semiproduct of the $[k(j){:}\ j$ in A] for each A of $FP(J)$. Then there must be a unique h: D$' \to$ D such that $q(A) \cdot h = k'(A)$ for each A, and so $s(j) \cdot h = k(j)$ for each j. Hence (D, s) is a limit of F in C. □

Combining this result with Theorem 6.3.1 parts (b) and (c), we get the following:

Corollary *A complete category C, which is well-powered, admits unique (E, M)-factorizations, with E as the class of extremal epimorphisms and M as the class of monomorphisms.*

For a complete category has pullbacks, equalizers, and intersections for sets of subobjects of an object.

EXERCISES AND REMARKS ON CHAPTER 6

Some typical examples of limits and completeness follow.

Example 1. The category S of sets. It is clear that in this category there exist arbitrary products and coproducts as well as equalizers and coequalizers. Hence

S is complete and cocomplete. Also S is well-powered and cowell-powered. So it admits unique (E^*, M) and unique (E, M^*)-factorizations, where E, M stand for the classes of epimorphisms and monomorphisms and E^*, M^* for the subclasses of extremal ones. But S is also balanced, so E^*, M^* coincide with E, M. The construction of the factorization in the proof of part (b) in Theorem 6.3.1 gives here, for a set map $f\colon A \to B$, a factorization $f = m \cdot e$ with m an inclusion map of a subset $f(A)$ in B and e a surjective map of A on $f(A)$. The dual construction would split f as $f = m' \cdot e'$ with e' a canonical surjection $A \to A/E$ and m an injective map of A/E in B, where $E = [(a, a')\colon f(a) = f(a')]$. These two factorizations are essentially equivalent.

Example 2. The category FA (or FDA) of F-algebras (FD-algebras) and their F-homomorphisms. This category has aribtrary products and equalizers; it is also well-powered. So it is complete and has unique (E^*, M)-factorizations. Also the category is balanced, so that $E^* = E$. Again two forms of factorizations (which are essentially equivalent) are possible: an epimorphism, followed by an inclusion monomorphism; or a canonical surjection on a quotient algebra by a congruence, followed by an injective homomorphism.

Example 3. An ordered set $(P, \leq)$. This is considered a small category; a nonnull set of elements has a product iff it has a greatest lower bound. So it is finitely complete iff it is a semilattice for g.l.b's, and finitely cocomplete iff it is a semilattice for l.u.b's. It is complete iff all subsets (nonnull subsets, if we are using only nonnull J's) have g.l.b.'s. Since any pair of objects (i, j) in P has at most one morphism from i to j, $(P, \leq)$ has trivially equalizers and coequalizers; and every morphism is both an epimorphism and a monomorphism, and an extremal epi- or monomorphism must be an isomorphism. We have (E, M)-factorizations trivially, an f written as $1 \cdot f$ or $f \cdot 1$.

Example 4. PO. This is the category of preordered sets and their monotone maps. It is a concrete category, has arbitrary products and coproducts, has equalizers and coequalizers, is well-powered and cowell-powered. So it is complete and cocomplete, having unique (E^*, M)- and unique (E, M^*)-factorizations. For an $f\colon (A, \leq) \to (B, \leq)$, the first factorization is of the form $f = m \cdot e$ where m is an inclusion morphism of a subset $f(A)$ of B with its relative preorder in $(B, \leq)$ and e a surjective monotone map of $(A, \leq)$ on this $(f(A), \leq)$. The other factorizes f as $m' \cdot e'$, with e' as a canonical surjective map of $(A, \leq)$ on a quotient preordered set $(A/E, \leq/E)$ where $E = [(a, a')\colon f(a) = f(a')]$ and m' is an injective monotone map of $(A/E, \leq/E)$ in $(B, \leq)$.

Example 5. T or SU. The categories of topological spaces and their continuous maps, as well as the category of semiuniform spaces and their uniform maps, have very similar characteristics. They have arbitrary products and coproducts, equalizers and coequalizers, and are well-powered and cowell-powered. So they are both complete and cocomplete, and have unique (E^*, M)- and unique (E, M^*)-factorizations. These look somewhat like the corresponding ones described in Example 4.

Exercises

1. Given a set of morphisms $[f_i\colon \mathrm{A}_i \to \mathrm{B}_i]$ of C and products $\mathrm{A} = P[\mathrm{A}_i]$, $\mathrm{B} = P[\mathrm{B}_i]$ with canonical projections $p_i\colon \mathrm{A} \to \mathrm{A}_i$, $q_i\colon \mathrm{B} \to \mathrm{B}_i$, there is a unique $f\colon \mathrm{A} \to \mathrm{B}$ such that $q_i \cdot f = f_i \cdot p_i$ for each i. This f we call the product of the $[f_i]$, and write $f = P[f_i]$. Verify the following results for $f = P[f_i]$:
   (a) $f = P[f_i] =$ the semiproduct $SP[f_i \cdot p_i]$;
   (b) the product morphism f is (i) a retraction, (ii) a section, (iii) an isomorphism, (iv) a monomorphism, (v) a regular monomorphism, or (vi) a constant morphism, when each of the f_i has the same property.

2. Given a functor $G\colon C \to D$ and a small category J, each functor $F\colon J \to C$ determines a functor $G \cdot F\colon J \to D$ and each predecessor (C, p) of F in C determines a predecessor $(F(\mathrm{C}), F \cdot p)$ of $G \cdot F$ in D [where $F \cdot p(j)$ is $F(p(j))$ for each j in $O(J)$]. We then say that G *preserves J-limits* if, for each functor $F\colon J \to C$, [(C, p) is a limit of F in C] implies [$(F(\mathrm{C}), F \cdot p)$ is a limit of $G \cdot F$ in D]; and G *reflects J-limits* if, for each functor $F\colon J \to C$, [(C, p) is a predecessor of F in C and $(F(\mathrm{C}), F \cdot p)$ is a limit of $G \cdot F$ in D] imply [(C, p) is a limit of F in C]. With these definitions we have the following results (prove these!):
   (i) Given the functor $G\colon C \to D$ and given that C is finitely complete, the following statements are equivalent: (a) G preserves J-limits for all finite categories J; (b) G preserves finite products and equalizers; (c) G preserves finite products and pullbacks. (G preserves pullbacks means just that it preserves J-limits for a particular form of J.)
   (ii) Given the functor $G\colon C \to D$ and given that C is complete, the following are equivalent: (a) G preserves J-limits for each small category J; (b) G preserves arbitrary products and equalizers; (c) G preserves arbitrary products and pullbacks; (d) G preserves finite products, equalizers, and inverse limits.
   (iii) If $G\colon C \to D$ is a faithful functor that reflects isomorphisms, then [C has J-limits and G preserves J-limits] implies [G reflects J-limits].
   (iv) A full and faithful functor $G\colon C \to D$ reflects J-limits for each J.

CHAPTER SEVEN

ADJOINT FUNCTORS

7.1 THE PATH CATEGORY

When we extend the semigroup of natural numbers to the group of integers or the multiplicative semigroup of positive integers to the multiplicative group of positive rationals, we are essentially going from a (commutative, cancellative) semigroup to a (commutative) group that is in some sense the nearest possible: there is a semigroup homomorphism (even isomorphism) of the semigroup in the group such that any other homomorphism of the semigroup in a group resolves as this homomorphism followed by a group homomorphism. A similar situation is encountered when we go from the incomplete uniform space of rationals to the complete uniform space of reals. All these suggest the following generalized formulation of a "shortest path."

Given two categories A, B and a functor F: $A \to B$, we now associate a new category, the *category of F-paths at an object* B of B as follows: a pair (f, A) is called an *F-path from* B *to A* if A is an object of A and f: $\mathrm{B} \to F(\mathrm{A})$ is a morphism in B. For two F-paths (f, A) and (f', A') from B to A, we say that g: $(f, \mathrm{A}) \to (f', \mathrm{A}')$ is a *path morphism* from the first path to the second if g: $\mathrm{A} \to \mathrm{A}'$ is a morphism in A such that $F(g) \cdot f = f'$. With composition of such path morphisms being essentially the composition of morphisms in A, and with $1_{(f,\mathrm{A})}$ taken as 1_{A}, these paths at B and path morphisms between these form a category $FP(\mathrm{B})$: the *category of F-paths at* B. If this category has an initial object (f^*, A^*), we call it a *shortest F-path at* B. Clearly such a path is unique up to an isomorphism in $FP(\mathrm{B})$.

Dually, for a given functor G: $B \to A$, a *G-copath* at an object A of A

is a pair (B, g) with B in $O(B)$ and g in $\hom_A(G(\mathrm{B}), \mathrm{A})$; these G-copaths form the objects of a category $GCP(\mathrm{A})$; if it has a final object (B^*, g^*) it is called a *shortest G-copath* at A. It would be unique up to an isomorphism.

Before a further analysis of these notions, we look at some examples.

Example 1. Given the forgetful functor G: $FA \to S$ or G: $FDA \to S$, there exist shortest G-paths at each object of S; for Lemmas 1.3.2 and 1.3.3 imply that $[j, [P(X, F), F]]$ and $[k, [P(X, F)/E^*, F]]$ are such shortest paths at X, where j is the inclusion map and k is $p \cdot j$, where p denotes the canonical surjective map of $P(X, F)$ on $P(X, F)/E^*$. We shall see how the construction of the free FD-algebra often involves a special selection of typical elements from the E^*-classes in the exercises at the end of this chapter.

Example 2. For the forgetful functor G: $PO \to S$, each object X of S has a shortest G-path $(1_X, (X, 1_X))$ and a shortest G-copath $((X, X \times X), 1_X)$. In the first case X has its finest preorder whereas in the second case X has its coarsest preorder.

Example 3. When G is the forgetful functor G: $T \to S$ or G: $SU \to S$, each X of $O(S)$ has a shortest G-path as well as a shortest G-copath; as in the last example, the morphism is 1_X for both, and the object is either X with its finest associated topology or semiuniformity (for the shortest path), or X with its coarsest topology or semiuniformity (for the copath).

When each object of B has a shortest G-path for a functor G: $A \to B$, as in Examples 1–3, we have another functor F: $B \to A$ with some characteristic properties that lead to the notion of adjoints.

7.2 ADJOINTNESS

Definition 7.2.1 If G: $A \to B$ and F: $B \to A$ are two functors, these determine two functors $[\hom_A \circ (F \times 1_A)]$ and $[\hom_B \circ (1_B \times G)]$, both from $B^{\mathrm{op}} \times A$ to S; the first is obtained by composing $F \times 1_A$: $B^{\mathrm{op}} \times A \to A^{\mathrm{op}} \times A$, with $\hom_A$: $A^{\mathrm{op}} \times A \to S$, and the second by composing $1_B \times G$: $B^{\mathrm{op}} \times A \to B^{\mathrm{op}} \times B$, with $\hom_B$: $B^{\mathrm{op}} \times B \to S$.

Definition 7.2.2 Given the functors G: $A \to B$ and F: $B \to A$, we say that F is a *left-adjoint* of G and that G is a *right-adjoint* of F if there exists a natural equivalence θ from $\hom_A \circ (F \times 1_A)$ to $\hom_B \circ (1_B \times G)$.

The connection between this adjointness notion and shortest paths or copaths is given in the next theorem.

Theorem 7.2.1 *(a) Given a functor G: $A \to B$, [each object* B *of B has a shortest G-path] iff [G has a left-adjoint functor F: $B \to A$].*
(b) Given a functor F: $B \to A$, [each object A *of A has a shortest F-copath] iff [F has a right-adjoint functor G: $A \to B$].*

PROOF. Clearly (a) and (b) are dual statements, and if we prove (a) then the proof of (b) follows by duality. So we shall only prove (a).

First we assume that each object B of B has a shortest G-path. We choose one such for each B and denote it by $(\eta(B), F(B))$. Thus F associates an object of A to each object of B, and $\eta(B)$: B $\to G(F(B))$ is a morphism in B. We want to define F for morphisms of B. Given b: B$' \to$ B from $M(B)$, we see that $(\eta(B) \cdot b, F(B))$ is a G-path at B$'$, while $(\eta(B'), F(B'))$ is a shortest G-path at B$'$, by assumption. Hence there must be a unique morphism g: $F(B') \to F(B)$ in A such that $G(g) \cdot \eta(B') = \eta(B) \cdot b$; we denote this unique g by $F(b)$. So $F(b)$: $F(B') \to F(B)$ when b: B$' \to$ B, and further, $GF(b) \cdot \eta(B') = \eta(B) \cdot b$. It is clear that $F(1_{B'}) = 1_{F(B')}$, and for a composite morphism $b \cdot b'$, where b: B$' \to$ B and b': B$'' \to$ B$'$, $F(b \cdot b') = F(b) \cdot F(b')$. For $F(b) \cdot F(b')$ is a morphism from $F(B'')$ to $F(B)$ such that $G[F(b) \cdot F(b')] \cdot \eta(B'') = GF(b) \cdot GF(b') \cdot \eta(B'') = GF(b) \cdot \eta(B') \cdot b' = \eta(B) \cdot b \cdot b'$, and this is the characteristic property that defines $F(b \cdot b')$. Hence we do have a functor F: $B \to A$. We also see, from the relation $GF(b) \cdot \eta(B') = \eta(B) \cdot b$ for any b: B$' \to$ B, that η: $1_B \to GF$ is a natural transformation.

To define the natural transformation (and equivalence) θ: $\hom_A \circ (F \times 1_A) \to \hom_B \circ (1_B \times G)$, for each object (B, A) from $B^{\mathrm{op}} \times A$ we have to define the morphism $\theta(B, A)$: $\hom_A(F(B), A) \to \hom_B(B, G(A))$. Given a morphism α from $\hom_A(F(B), A)$, we set $\theta(B, A)(\alpha) = G(\alpha) \cdot \eta(B)$ (see Figure 7.1). This evidently belongs to $\hom_B(B, G(A))$. For any β from $\hom_B(B, G(A))$, since (β, A) would be a G-path at B while $(\eta(B), F(B))$ is a shortest G-path at B, there is a unique morphism α in $\hom_A(F(B), A)$ for which $G(\alpha) \cdot \eta(B) = \beta$; this means that $\theta(B, A)$ is both surjective and injective from $\hom_A(F(B), A)$ to $\hom_B(B, G(A))$. Hence it now suffices to check that θ is a natural transformation in order to ensure that it is a natural equivalence (since in S bijections are the isomorphisms). Given then a pair of objects (B, A) and (B$'$, A$'$) of $B^{\mathrm{op}} \times A$ and a morphism (b, a) from the first to the second (so that b: B$' \to$ B in B and a: A $\to$ A$'$ in A), we have to check the commutativity condition: for any α in $\hom_A(F(B), A)$, $\theta(B', A')[a \cdot \alpha \cdot F(b)] = G(a) \cdot \theta(B, A)(\alpha) \cdot b$ (see Figure 7.2). That is, we have to prove the equality $G(a \cdot \alpha \cdot F(b)) \cdot \eta(B') = G(a) \cdot (G(\alpha) \cdot \eta(B)) \cdot b$, or $G(a) \cdot G(\alpha) \cdot GF(b) \cdot \eta(B') = G(a) \cdot G(\alpha) \cdot \eta(B) \cdot b$, which follows from our earlier equality $GF(b) \cdot \eta(B') = \eta(B) \cdot b$.

To prove the converse result assume now that there is a natural equivalence θ from $\hom_A \circ (F \times 1_A)$ to $\hom_B \circ (1_B \times G)$. This means that for

F(b)
F(B)
F(B′)
α
α′
A
a
A′
A
F
G
GF(b)
GF(B)
GF(B′)
G(α)
η(B′)
G(α′)
B
η(B)
G(A)
G(a)
G(A′)
θ(B, A)(α)
θ(B′, A′)(α′)
B
b
B′

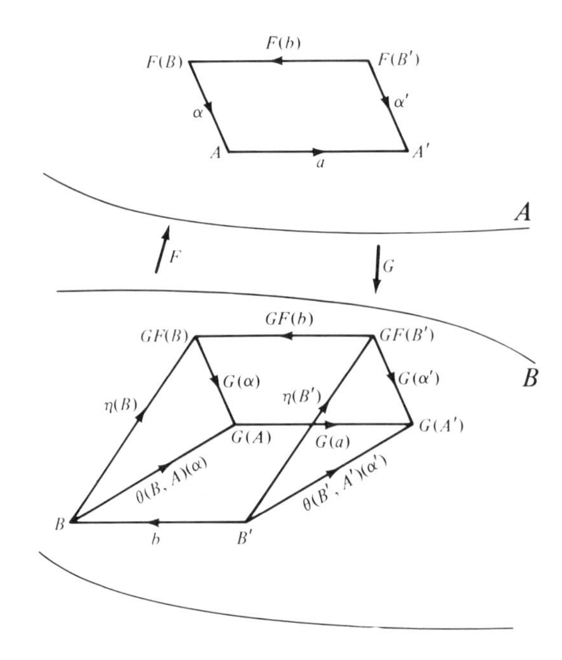

Figure 7.1

each object (B, A) from $B^{op} \times A$, θ(B, A) is a bijective map of $\hom_A(F(\mathrm{B}), \mathrm{A})$ in $\hom_B(\mathrm{B}, G(\mathrm{A}))$. Taking in particular A = F(B) for a certain B from $O(B)$, θ(B, F(B)) is a bijection of $\hom_A(F(\mathrm{B}), F(\mathrm{B}))$ in $\hom_B(\mathrm{B}, GF(\mathrm{B}))$. Corresponding to the element $1_{F(\mathrm{B})}$ of the first set we have an element $\theta(\mathrm{B}, F(\mathrm{B}))(1_{F(\mathrm{B})})$ in $\hom_B(\mathrm{B}, GF(\mathrm{B}))$. We denote this morphism by η(B). Then we shall check that (η(B), F(B)) is a shortest G-path at the object B of B. It surely is a G-path at B. Given any G-path

Figure 7.2

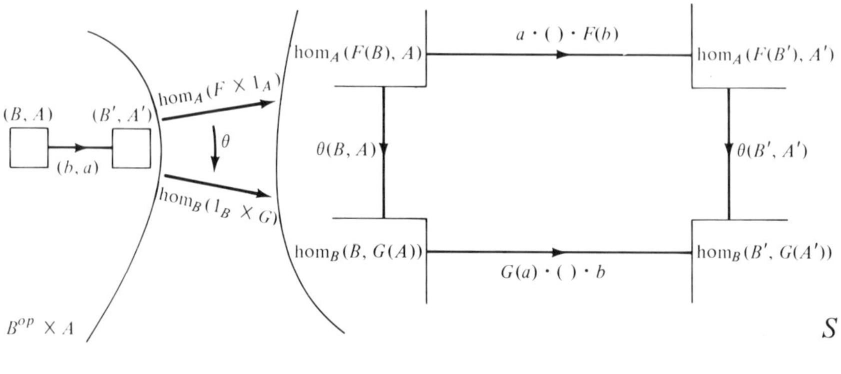

(B, A)
(B′, A′)
(b, a)
hom_A(F × 1_A)
θ
hom_B(1_B × G)
B^op × A
hom_A(F(B), A)
a · ( ) · F(b)
hom_A(F(B′), A′)
θ(B, A)
θ(B′, A′)
hom_B(B, G(A))
G(a) · ( ) · b
hom_B(B′, G(A′))
S

(f, A) at B, so that f: B $\rightarrow F(A)$ in B, we have a unique g: $F(B) \rightarrow$ A in A such that $\theta(B, A)(g) = f$. Given any morphism g in $\hom_A(F(B), A)$, we have a morphism $(1_B, g)$ in $B^{op} \times A$ from $(B, F(B))$ to (B, A). Corresponding to this morphism in $B^{op} \times A$, θ provides a commutative rectangle (in S): $[\hom_B \circ (1_B \times G)](1, g) \cdot \theta(B, F(B)) = \theta(B, A) \cdot [\hom_A \circ (F \times 1_A)](1, g)$ (see Figure 7.3). The two morphisms here take $1_{F(B)}$ of $\hom_A(F(B), F(B))$ to $G(g) \cdot \eta(B) \cdot 1_B$ and to $[\theta(B, A)](g \cdot 1_{F(B)})$; thus $G(g) \cdot \eta(B) = \theta(B, A)(g)$ for each g of $\hom_A(F(B), A)$. Thus our earlier assertion that for any f of $\hom_B(B, G(A))$ (or for any G-path (f, A) at B) there is a unique g in $\hom_A(F(B), A)$ such that $\theta(B, A)(g) = f$ can now be used to deduce that for each G-path (f, A) at B there is a unique g in $\hom_A(F(B), A)$ such that $G(g) \cdot \eta(B) = f$. This is just the condition we want in order to ensure that $(\eta(B), F(B))$ is indeed a shortest G-path at B. □

In proving part (a) of Theorem 7.2.1 we chose one of many possible shortest G-paths $(\eta(B), F(B))$ and used this $F(B)$ to construct a left-adjoint F for G. What would happen if we were to change the choice here? Had we chosen another, say $(\eta'(B), F'(B))$, for each B of $O(B)$, then the F' constructed as a left-adjoint of G would be naturally equivalent with F; this is as much freedom as there can be in finding a left-adjoint of G. These are the main conclusions of the following lemma.

Lemma 7.2.1 *Suppose F, F′ are two functors from B to A and G, G′ are two functors from A to B. Then we have the following:*

(a) *given F is a left-adjoint of G, [F′ is a left-adjoint of G] iff [F and F′ are naturally equivalent];*
(b) *given, G is a right-adjoint of F, [G′ is a right-adjoint of F] iff [G and G′ are naturally equivalent];*
(c) *given F is a left-adjoint of G, [F and F′ are naturally equivalent and G and G′ are naturally equivalent] implies that [F′ is a left-adjoint of G′].*

Figure 7.3

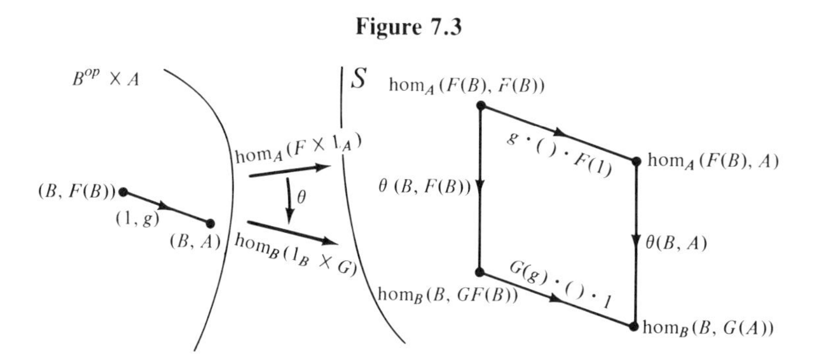

PROOF. Clearly, result (b) is the dual of result (a), so can be proved by duality once (a) is proved; while (c) follows by using (a) and (b) together. So we have only to prove (a).

Assume then that F is a left-adjoint of G and that F, F' are naturally equivalent. As before, we denote by θ a natural equivalence θ: $\hom_A \circ (F \times 1_A) \to \hom_B \circ (1_B \times G)$, so that for a typical (b, a) in $\hom_{B^{op} \times A}$ [(B, A), (B′, A′)], we have for each h in $\hom_A(F(\mathrm{B}), \mathrm{A})$ a commutativity relation $\theta(\mathrm{B}', \mathrm{A}')[a \cdot h \cdot F(b)] = G(a) \cdot [\theta(\mathrm{B}, \mathrm{A})(h)] \cdot b$; and we assume that λ: $F \to F'$ is a natural equivalence, so that for a typical b: $\mathrm{B}' \to \mathrm{B}$ of B we have the commutative relation $\lambda(\mathrm{B}) \cdot F(b) = F'(b) \cdot \lambda(\mathrm{B}')$. To show that F' is also a left-adjoint of G, we define a mapping $\theta'(\mathrm{B}, \mathrm{A})$: $\hom_A(F'(\mathrm{B}), \mathrm{A}) \to \hom_B(\mathrm{B}, G(\mathrm{A}))$ for each object (B, A) from $B^{op} \times A$ as follows: for any h' in $\hom_A(F'(\mathrm{B}), \mathrm{A})$, $\theta'(\mathrm{B}, \mathrm{A})(h') = \theta(\mathrm{B}, \mathrm{A})[h' \cdot \lambda(\mathrm{B})]$, which is surely an element of $\hom_B(\mathrm{B}, G(\mathrm{A}))$. Since $\theta(\mathrm{B}, \mathrm{A})$ is a bijective map and $\lambda(\mathrm{B})$ is an isomorphism in A, it is not hard to see that $\theta'(\mathrm{B}, \mathrm{A})$ is also a bijective map. So if we show that this θ' is a natural transformation from $\hom(F' \times 1_A)$ to $\hom(1_B \times G)$, then it would be a natural equivalence, and so F' would be a left-adjoint of G. To see the naturalness of θ', note that if (b, a) is a morphism in $B^{op} \times A$ as before, from (B, A) to (B′, A′), then for any h' of $\hom_A(F'(\mathrm{B}), \mathrm{A})$ we have the relations:

$$\begin{aligned} \theta'(\mathrm{B}', \mathrm{A}')[a \cdot h' \cdot F'(b)] &= \theta(\mathrm{B}', \mathrm{A}')[a \cdot h' \cdot F'(b) \cdot \lambda(\mathrm{B}')] \\ &= \theta(\mathrm{B}', \mathrm{A}')[a \cdot h' \cdot \lambda(\mathrm{B}) \cdot F(b)] \\ &= G(a) \cdot [\theta(\mathrm{B}, \mathrm{A})(h' \cdot \lambda(\mathrm{B})] \cdot b \\ &= G(a) \cdot \theta'(\mathrm{B}, \mathrm{A})(h') \cdot b, \end{aligned}$$

which is the required commutativity relation.

For the reverse part of (a) we assume that both F and F' are left-adjoints of G, and we prove that there is a natural equivalence from F' to F. As in the proof of part (b) of Theorem 7.2.1 we have a shortest G-path $(\eta(\mathrm{B}), F(\mathrm{B}))$ at B and another shortest G-path $(\eta'(\mathrm{B}), F'(\mathrm{B}))$, also at B, with η: $1_B \to G \cdot F$ and η': $1_B \to G \cdot F'$ being natural transformations. Hence, by the definition of a shortest G-path at B, there must be an isomorphism $\lambda(\mathrm{B})$: $F'(\mathrm{B}) \to F(\mathrm{B})$ (which is even unique) such that $\eta(\mathrm{B}) = G(\lambda(\mathrm{B})) \cdot \eta'(\mathrm{B})$. (See Figure 7.4.) Since $(\eta'(\mathrm{B}'), F'(\mathrm{B}'))$ is a shortest G-path at B′, while $(\eta(\mathrm{B}) \cdot b, F(\mathrm{B}))$ is also a G-path at B′, there is a unique morphism m from $F'(\mathrm{B}')$ to $F(\mathrm{B})$ in A such that $G(m) \cdot \eta'(\mathrm{B}') = \eta(\mathrm{B}) \cdot b$. Now for $F(b) \cdot \lambda(\mathrm{B}')$ we have $G(F(b) \cdot \lambda(\mathrm{B}')) \cdot \eta'(\mathrm{B}') = G \cdot F(b) \cdot G(\lambda(\mathrm{B}')) \cdot \eta'(\mathrm{B}') = G \cdot F(b) \cdot \eta(\mathrm{B}') = \eta(\mathrm{B}) \cdot b$, since η is natural; and for $\lambda(\mathrm{B}) \cdot F'(b)$ also we have $G(\lambda(\mathrm{B}) \cdot F'(b)) \cdot \eta'(\mathrm{B}') = G(\lambda(\mathrm{B})) \cdot G \cdot F'(b) \cdot \eta'(\mathrm{B}') = G(\lambda(\mathrm{B})) \cdot \eta'(\mathrm{B}) \cdot b$ (since η' is natural)

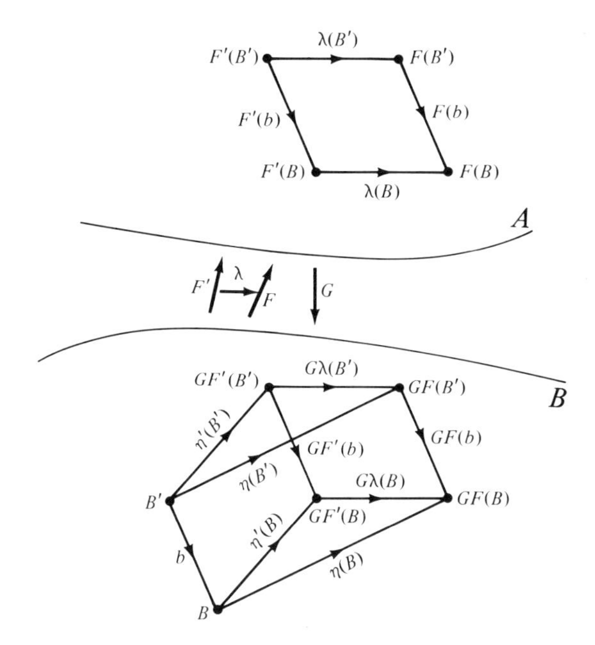

Figure 7.4

$= \eta(B) \cdot b$. Hence, by our definition and the uniqueness of m, we get $F(b) \cdot \lambda(B') = \lambda(B) \cdot F'(b)$, which is the required commutativity relation to prove that $\lambda: F' \to F$ is a natural transformation. This completes the proof of (a) and so of the lemma. □

7.3 NEAR-EQUIVALENCE AND ADJOINTNESS

In Section 5.2 we defined the equivalence of categories and an equivalence scheme. Near-equivalence, a slight generalization of this notion, is closely related to adjointness.

Definition 7.3.1 The categories A and B are called *near-equivalent*, and $(\eta, \nu; F, G)$ is called a *near-equivalence scheme for* (A, B), if there exist functors $F: A \to B$, $G: B \to A$ and also two natural transformations $\eta: 1_B \to (F \cdot G)$ and $\nu: (G \cdot F) \to 1_A$, with the further properties:

(a) $(\nu \# 1_G) \circ (1_G \# \eta) = 1_G$[that is, for all $B \in O(B)$, $\nu(G(B)) \cdot G(\eta(B)) = 1_{G(B)}$],

(b) $(1_F \# \nu) \circ (\eta \# 1_F) = 1_F$ [that is, for all $A \in O(A)$, $F(\nu(A)) \cdot \eta(F(A)) = 1_{F(A)}$].

We observe that in Theorem 5.2.1, part (d), we showed that a functor that is faithful, full, and dense from a category A to a category B determines an equilvalence scheme, which is also a near-equivalence scheme according to Definition 7.3.1.

We now relate near-equivalence to adjoints.

Theorem 7.3.1 *Categories A and B are near-equivalent iff there exist functors $F\colon A \to B$ and $G\colon B \to A$ such that G is a left-adjoint of F.*

PROOF. First assume that there exist functors $F\colon A \to B$ and $G\colon B \to A$ such that G is a left-adjoint of F. Using the proof of the "if" part of Theorem 7.2.1, with the difference that the roles of F and G are now interchanged, we have a natural transformation η from 1_B to $F \cdot G$, and dually $\nu\colon G \cdot F \to 1_A$. We know that η and ν are defined in terms of the natural equivalence $\theta\colon \hom_A \circ (G \times 1_A) \to \hom_B \circ (1_B \times F)$ by setting $\eta(\mathrm{B}) = \theta(\mathrm{B}, G(\mathrm{B}))(1_{G(\mathrm{B})})$ for each B in $O(B)$ and $\nu(\mathrm{A}) = [\theta(F(\mathrm{A}), \mathrm{A})]^{\mathrm{r}}(1_{F(A)})$ for each A in $O(A)$. Considering the morphism $(\eta(\mathrm{B}), 1_{G(\mathrm{B})})\colon (FG(\mathrm{B}), G(\mathrm{B})) \to (\mathrm{B}, G(\mathrm{B}))$ in $B^{\mathrm{op}} \times A$, we see that θ gives a commutativity relation of set maps $\hom(1_B \times F)(\eta(\mathrm{B}), 1_{G(\mathrm{B})}) \circ \theta(FG(\mathrm{B}), G(\mathrm{B})) = \theta(\mathrm{B}, G(\mathrm{B})) \circ \hom(G \times 1_A)\ (\eta(\mathrm{B}), 1_{G(\mathrm{B})})$. Applying these equal maps to the element $\nu(G(\mathrm{B}))$ in $\hom_A(GFG(\mathrm{B}), G(\mathrm{B}))$, we get $1_{FG(\mathrm{B})} \cdot [\theta(FG(\mathrm{B}), G(\mathrm{B}))](\nu(G(\mathrm{B}))) \cdot \eta(\mathrm{B}) = \theta(\mathrm{B}, G(\mathrm{B}))[1_{G(\mathrm{B})} \cdot \nu(G(\mathrm{B})) \cdot G(\eta(\mathrm{B}))]$; but the left-hand side is just $1_{FG(\mathrm{B})} \cdot 1_{FG(\mathrm{B})} \cdot \eta(\mathrm{B}) = \eta(\mathrm{B}) = \theta(\mathrm{B}, (G(\mathrm{B})) \cdot (1_{G(\mathrm{B})})$. Since $\theta(\mathrm{B}, G(\mathrm{B}))$ is a bijective map, these then give the equality of $1_{G(\mathrm{B})}$ and $\nu(G(\mathrm{B})) \cdot G(\eta(\mathrm{B}))$. Thus (a) is proved; (b) can be proved similarly, using the same θ for the morphism $(1_{F(\mathrm{A})}, \nu(\mathrm{A}))\colon (F(\mathrm{A}), GF(\mathrm{A})) \to (F(\mathrm{A}), \mathrm{A})$ in $B^{\mathrm{op}} \times A$. Hence we have shown that $(\eta, \nu; F, G)$ is a near-equivalence scheme for A, B, so that A, B are near-equivalent.

For proving the converse, let us assume that there is a near-equivalence scheme $(\eta, \nu; F, G)$ for A, B. To prove then that G is a left-adjoint of F, we show that for each object B of B, $(\eta(\mathrm{B}), G(\mathrm{B}))$ is a shortest G-path at B. Since $\eta(\mathrm{B})\colon \mathrm{B} \to FG(\mathrm{B})$, surely $(\eta(\mathrm{B}), G(\mathrm{B}))$ is a G-path at B. Let (f, A) be any other path at B, so that $f\colon \mathrm{B} \to F(\mathrm{A})$. Then there is a morphism, namely, $\nu(\mathrm{A}) \cdot G(f)$ in $\hom_A(G(\mathrm{B}), \mathrm{A})$ such that $F(\nu(\mathrm{A}) \cdot G(f)) \cdot \eta(\mathrm{B}) = F\nu(\mathrm{A}) \cdot FG(f) \cdot \eta(\mathrm{B}) = F\nu(\mathrm{A})\, \eta(F(\mathrm{A})) \cdot f$ (since η is natural) $= 1_{F(\mathrm{A})} \cdot f$ [by (b)] $= f$. Moreover, if for any m' in $\hom_A(G(\mathrm{B}), \mathrm{A})$ it is true that $F(m') \cdot \eta(\mathrm{B}) = f$, then $GF(m') \cdot G(\eta(\mathrm{B})) = G(f)$ and so $\nu(\mathrm{A}) \cdot G(f) = \nu(\mathrm{A}) \cdot GF(m') \cdot G(\eta(\mathrm{B})) = m' \cdot \nu(G(\mathrm{B})) \cdot G(\eta(\mathrm{B}))$ (since ν is natural) $= m' \cdot 1_{G(\mathrm{B})}$ by (a). Hence $m' = \nu(\mathrm{A}) \cdot G(f)$ for such an m'. These prove that indeed $(\eta(\mathrm{B}), G(\mathrm{B}))$ is a shortest G-path at B. Hence F is a left-adjoint of G [as in the proof of Theorem 7.2.1, part (a)]. (See Figure 7.5.) □

It could happen,as we shall see later when we look at examples, that we have a situation between equivalence and near-cquivalence. We formulate definitions for two such cases.

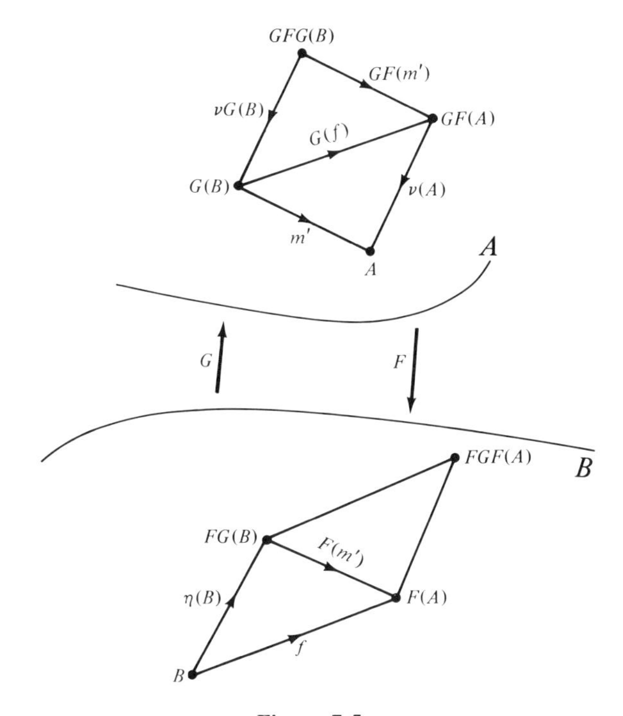

Figure 7.5

Definition 7.3.2 When $(\eta, \nu; F, G)$ is a near-equivalence scheme for A, B we call (i) G a *retractor* of B in A if ν is a natural equivalence; (ii) F a *coretractor* of A in B if η is a natural equivalence.

(When G is a retractor of B in A and further A is a full subcategory of B with F as the inclusion functor, G becomes the "reflection" of B in A in the sense of Herrlich and Strecker; see [15], Chapter X.)

7.4 COMPOSING AND RESOLVING SHORTEST PATHS OR ADJOINTS

Given three categories A, B, and C and functors F: $A \to B$, P: $B \to C$, when F and P have left-adjoints G: $B \to A$ and Q: $C \to B$, is $G \cdot Q$ a left-adjoint of $P \cdot F$? It is; and more can be proved, as the next lemma and theorem show.

Lemma 7.4.1 *Given the functors F: $A \to B$ and P: $B \to C$, we have the following:*

(a) if for a C *in $O(C)$ there is a shortest P-path $(\lambda(\mathrm{C}), \mathrm{B})$ and for this* B *of $O(B)$ there is a shortest F-path $(\eta(\mathrm{B}), \mathrm{A})$, then $(P(\eta(\mathrm{B})) \cdot \lambda(\mathrm{C}), \mathrm{A})$ is a shortest $(P \cdot F)$-path at* C*;*

(*b*) *if for a* C *of* $O(C)$ *there is a shortest* P*-path* $(\lambda(C), B)$ *and a shortest* $(P \cdot F)$*-path* $(\mu(C), A)$, *then there is a shortest* F*-path* $(\eta(B), A)$ *for* B *such that* $\mu(C) = P(\eta(B)) \cdot \lambda(C)$.

PROOF. (See Figure 7.6.) (a) It is clear that under the hypothesis we have a $(P \cdot F)$-path $(P(\eta(B)) \cdot \lambda(C), A)$ at C. If (h, A') is another $(P \cdot F)$-path at C, so that f: $C \to P \cdot F(A')$ in C, then $(h, F(A'))$ is a P-path at C. Since $(\lambda(C), B)$ is a shortest P-path at C, there is a unique morphism θ: $B \to F(A')$ in B such that $h = P(\theta) \cdot \lambda(C)$. Then because (θ, A') is an F-path at B, while $(\eta(B), A)$ is a shortest such F-path, there is a unique morphism k: $A \to A'$ in A such that $\theta = F(k) \cdot \eta(B)$; hence, we also have $PF(k) \cdot P(\eta(B)) \cdot \lambda(C) = P(\theta) \cdot \lambda(C) = h$. To show that this morphism k is unique, let k': $A \to A'$ also satisfy the condition $h = PF(k') \cdot P(\eta(B)) \cdot \lambda(C) = P(F(k') \cdot \eta(B)) \cdot \lambda(C)$; the asserted uniqueness of θ implies that $F(k')\eta(B) = \theta$, and then the asserted uniqueness of k implies that $k' = k$. Thus $(P(\eta(B)) \cdot \lambda(C), A)$ is a shortest $(P \cdot F)$-path at C.

(b) Now we assume that $(\lambda(C), B)$ and $(\mu(C), A)$ give a shortest P-path and a shortest $P \cdot F$-path at C in $O(C)$; since $(\mu(C), F(A))$ is a P-path at C also, there is a unique morphism $\eta(B)$: $B \to F(A)$ in B such that $\mu(C) = P(\eta(B)) \cdot \lambda(C)$. Then it is clear that $(\eta(B), A)$ is an F-path at B; if (θ, A') is any other F-path at B, $(P(\theta) \cdot \lambda(C), A')$ is a $(P \cdot F)$-path at C; since $(\mu(C), A)$ is a shortest $(P \cdot F)$-path at C, there must be a unique morphism k: $A \to A'$ in A such that $P(\theta) \cdot \lambda(C) = (PF)(k) \cdot \mu(C) = (P \cdot F)(k) \cdot P(\eta(B)) \cdot \lambda(C) = P[F(k) \cdot \eta(B)] \cdot \lambda(C)$. Calling this morphism h in C from C to $PF(A')$, since $(h, F(A'))$ is a P-path at C and $(\lambda(C), B)$ is a shortest P-path at C, there must be a unique morphism g: $B \to F(A')$ for which $P(g) \cdot \lambda(C) = h$; since both θ and $F(k) \cdot \eta(B)$ are possible values for this g, they must be equal. If $\theta = F(k') \cdot \eta(B)$ for any other k': $A \to A'$, we would also get $PF(k') \cdot P(\eta(B)) \cdot \lambda(C) = PF(k) \cdot P(\eta(B)) \cdot \lambda(C)$ or $PF(k')\mu(C) = PF(k) \cdot \mu(C)$; but then [$(\mu(C), A)$ is a shortest PF-path] would imply that $k' = k$. Thus there is a unique k: $A \to A'$ such that $F(k) \cdot \eta(B) = \theta$. This proves that $(\eta(B), A)$ is a shortest F-path at B, as required. □

From this lemma we get the following theorem:

Theorem 7.4.1 (*a*) *If* $F: A \to B$ *has a left-adjoint* $G: B \to A$, *and* $P: B \to C$ *has a left-adjoint* $Q: C \to B$, *then* $P \cdot F: A \to C$ *has a left-adjoint* $G \cdot Q: C \to A$.

(*b*) *Given the functors* $F: A \to B$ *and* $P: B \to C$, *if* P *has a left-adjoint and retractor* Q *of* C *in* B *and* $(P \cdot F)$ *has a left-adjoint* $R: C \to A$, *then* F *has a left-adjoint* G *such that* $G \cdot Q$ *is naturally equivalent to* R.

PROOF. (a) When G, Q are left-adjoints of F and P, respectively, we can define η: $1_B \to F \cdot G$ and λ: $1_C \to P \cdot Q$, two natural transformations, such that (η(B), G(B)) is a shortest F-path at B for each B of $O(B)$ and (λ(C), Q(C)) is a shortest P-path at C for each C of $O(C)$, as in the proof of Theorem 7.2.1. Then from part (a) of Lemma 7.4.1 we have $P(\eta(Q(\mathrm{C}))) \cdot \lambda(\mathrm{C}), G(Q(\mathrm{C}))$) is a shortest ($P \cdot F$)-path at C for each C of $O(C)$. As in Theorem 7.2.1, we can then use this $G \cdot Q$: $O(C) \to O(A)$ to define a map for the objects of C and extend it to a functor R from C to A that would be a left-adjoint of $P \cdot F$; the way the R is defined for a morphism t: C $\to$ C$'$ in C is by setting $R(t)$ equal to that unique morphism g: $G \cdot Q(\mathrm{C}) \to G \cdot Q(\mathrm{C}')$ for which $P \cdot F(g) \cdot P(\eta(Q(\mathrm{C}))) \cdot \lambda(\mathrm{C}) = P(\eta(Q(\mathrm{C}'))) \cdot \lambda(\mathrm{C}') \cdot t$; since η, λ are natural transformations, we have $F \cdot G \cdot (Q(t)) \cdot \eta(Q(\mathrm{C})) = \eta(Q(\mathrm{C}')) \cdot Q(\mathrm{t})$, $P \cdot Q(t) \cdot \lambda(\mathrm{C}) = \lambda(\mathrm{C}') \cdot t$. Hence, $P \cdot F(G \cdot Q(t)) \cdot P(\eta(Q(\mathrm{C}))) \cdot \lambda(\mathrm{C}) = P[FGQ(t) \cdot \eta(Q(\mathrm{C}))] \cdot \lambda(\mathrm{C}) = P(\eta(Q(\mathrm{C}')))PQ(t) \cdot \lambda(\mathrm{C}) = P(\eta(Q(\mathrm{C}'))) \cdot \lambda(\mathrm{C}') \cdot t$. This shows that $G \cdot Q(t) = R(t)$ for each t of $M(C)$, so that $G \cdot Q$ is the left-adjoint of $P \cdot F$ constructed from the object function.

(b) If Q: $C \to B$ is a left-adjoint and retractor of P: $B \to C$, and R: $C \to A$ is a left-adjoint of $P \cdot F$: $A \to C$ (where F: $B \to C$ is a given functor), we can find natural transformations ν: $Q \cdot P \to 1_B$, λ: $1_C \to P \cdot Q$, and μ: $1_C \to (P \cdot F) \cdot R$ such that (i) ν is a natural equivalence, (ii) (λ(C), Q(C)), and (μ(C), R(C)) are shortest P- and ($P \cdot F$)-paths, respectively, at C for each object C of C. From the fact that ν is a natural equivalence it is easy to see that the functor P must be a faithful functor. Starting now with an object B of B, we have a near-equivalence scheme (λ, ν; P, Q) for (B, C); so, as in Theorem 7.3.1, we have $P(\nu(\mathrm{B})) \cdot \lambda(P(\mathrm{B})) = 1_{P(\mathrm{B})}$. Since we also know that ν(B) is an isomorphism in B, it has an inverse $(\nu(\mathrm{B}))^{\mathrm{r}}$ and P maps these into an isomorphism and inverse in C. Thus from the last equality we get $\lambda(P(\mathrm{B})) = [P(\nu(\mathrm{B}))]^{\mathrm{r}} = P[\nu(\mathrm{B})^{\mathrm{r}}]$. Now we have a P-path $[\mu(P(\mathrm{B})), F \cdot R(P(\mathrm{B}))]$ at P(B) and a shortest P-path $[\lambda(P(\mathrm{B})), Q \cdot P(\mathrm{B})]$, also at P(B). Hence there is a unique morphism δ(B): $Q \cdot P(\mathrm{B}) \to F \cdot R \cdot (P(\mathrm{B}))$ such that

$$P(\delta(\mathrm{B})) \cdot \lambda(P(\mathrm{B})) = \mu(P(\mathrm{B})) \tag{7.4.1}$$

See Figure 7.6. We define η': $1_B \to FRP$ by $\eta'(\mathrm{B}) = \delta(\mathrm{B}) \cdot [\nu(\mathrm{B})]^{\mathrm{r}}$ for each B of $O(B)$. So (η'(B), RP(B)) is an F-path at B. We claim that it is a shortest F-path. Let (θ, A$'$) be some F-path at B; then ($P(\theta)$, A$'$) is a $P \cdot F$-path, while ($\mu(P(\mathrm{B}))$, RP(B)) is a shortest $P \cdot F$-path at P(B). Hence there is a unique g: $RP(\mathrm{B}) \to$ A$'$ in A such that

$$PF(g) \cdot \mu(P(\mathrm{B})) = P(\theta). \tag{7.4.2}$$

Using both (7.4.1) and (7.4.2) we get $P(F(g)) \cdot P(\delta(\mathrm{B})) \cdot \lambda(P(\mathrm{B})) = P(\theta)$; hence it follows that $P[F(g) \cdot \eta'(\mathrm{B})] = P[F(g) \cdot \delta(\mathrm{B}) \cdot (\nu(\mathrm{B}))^{\mathrm{r}}] = PF(g) \cdot P\delta(\mathrm{B}) \cdot P[\nu(\mathrm{B})]^{\mathrm{r}} = PF(g) \cdot P\delta(\mathrm{B}) \cdot \lambda(P(\mathrm{B})) = P(\theta)$. But we already observed that P is faithful; so we deduce that $F(g) \cdot \eta'(\mathrm{B}) = \theta$. If

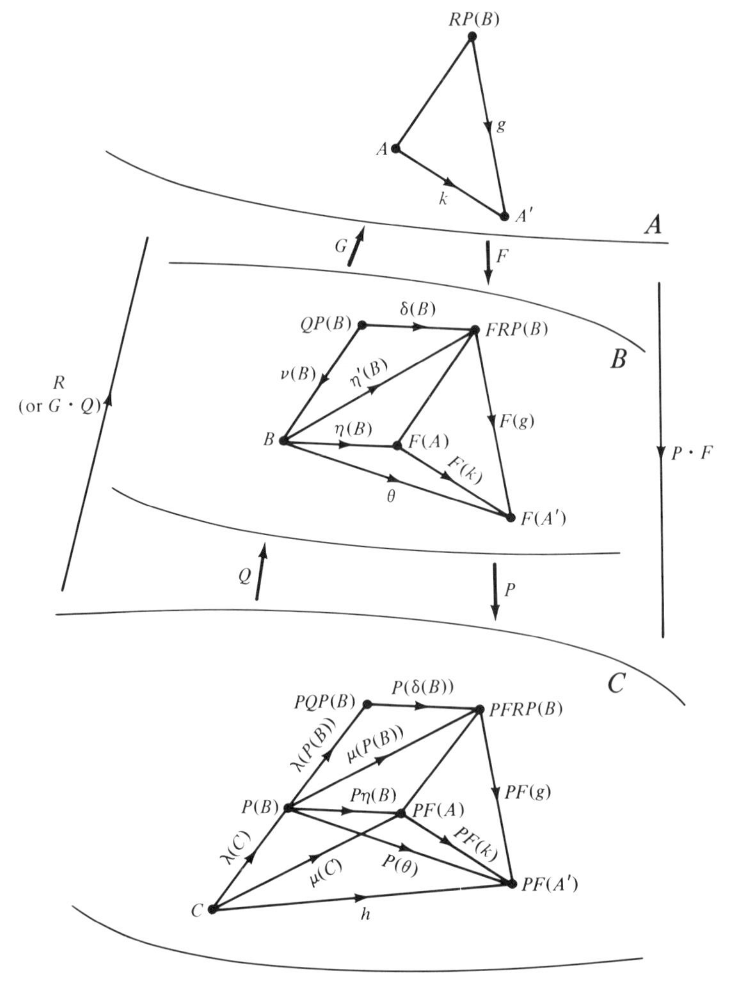

Figure 7.6

any other g': $RP(\mathrm{B}) \to \mathrm{A}$ satisfies $F(g') \cdot \eta'(\mathrm{B}) = \theta$, then $PF(g') \cdot P(\delta(\mathrm{B})) \cdot \lambda(P(\mathrm{B})) = PF(g') \cdot P(\delta(\mathrm{B})) \cdot P[\nu(\mathrm{B})^r] = PF(g') \cdot P\eta'(\mathrm{B}) = P(\theta)$, or $PF(g') \cdot \mu(P(\mathrm{B})) = P(\theta)$. The uniqueness asserted for g in (7.4.2) then gives $g' = g$. Thus we have proved that $(\eta'(\mathrm{B}), RP(\mathrm{B}))$ is indeed a shortest F-path at B. Since one such exists for each B of $O(B)$, F has a left-adjoint G. Then by part (a) proved earlier we deduce that $(G \cdot Q)$ is a left-adjoint of $(P \cdot F)$. Since by assumption R too is a left-adjoint of $(P \cdot F)$, it follows that R and $G \cdot Q$ are naturally equivalent. □

Regarding the existence of an adjoint for a given functor, we have the so-called *adjoint functor theorems*.

7.5 ADJOINT FUNCTOR THEOREMS

Lemma 7.5.1 *(a) If the functor $G: A \to B$ has a left-adjoint $F: B \to A$, then G preserves all existing limits in A (and dually, F preserves all colimits in B).*

(b) If the category A is complete, and the functor $G: A \to B$ preserves all limits in A, then an object B *of B has a shortest G-path provided the following "solution set condition" (ssc) is true: (ssc for* B *relative to G) there is a set $[h(k), \mathrm{A}(k)]$, k in K, of G-paths at* B *such that for any G-path (h, A) at* B *there is some k in K and a morphism $\theta: \mathrm{A}(k) \to \mathrm{A}$ such that $h = G(\theta) \cdot h(k)$.*

PROOF. (a) Suppose that $G: A \to B$ has a left-adjoint $F: B \to A$. Let J be any small category, $t: j \to k$ be a typical morphism in J. Let $D: J \to A$ be a functor. If D has a limit (A^*, p^*) in A, we have to show that $(G(\mathrm{A}^*), Gp^*)$ is a limit of $G \cdot D$ in B. (See Figure 7.7.)

For each j of $O(J)$, $Gp^*(j): G(\mathrm{A}^*) \to GD(j)$ is a morphism in B, and for the typical morphism $t: j \to k$ of J we have $GD(t) \cdot Gp^*(j) = Gp^*(k)$, since $D(t) \cdot p^*(j) = p^*(k)$ in A. Thus $(G(\mathrm{A}^*), Gp^*)$ is a predecessor of GD in B. If (B, q) is any predecessor of GD in B, we have $q(j); \mathrm{B} \to GD(j)$ in B for each j of $O(J)$ and $GD(t) \cdot q(j) = q(k)$ for $t: j \to k$ in J. Hence we have $Fq(j): F(\mathrm{B}) \to FGD(j)$ in A, and $FGD(t) \cdot Fq(j) = Fq(k)$ for t of $M(J)$. Because we have a natural transformation $\nu: FG \to 1_A$, $D(t) \cdot \nu(D(j)) = \nu(D(k)) \cdot FGD(t)$. This combined with the last relation gives $D(t) \cdot \nu(D(j)) \cdot Fq(j) = \nu(D(k)) \cdot FGD(t) \cdot Fq(j) = \nu(D(k)) \cdot Fq(k)$. If we set $q^*(j) = \nu(D(j)) \cdot Fq(j)$, $q^*(t) = D(t)$ for j, t of J, it is not hard to see that q^* is a functor from J to A, and the last result shows that $(F(\mathrm{B}), q^*)$ is a predecessor of D. Since (A^*, p^*) is a limit of D, there must be a unique morphism $m: F(\mathrm{B}) \to \mathrm{A}^*$ such that $p^*(j) \cdot m = q^*(j) = \nu(D(j)) \cdot Fq(j)$ for each j of $O(J)$. Let $r = G(m) \cdot \eta(\mathrm{B})$, where $\eta: 1_B \to GF$ is the natural transformation defined as before from the adjointness of F to G. Then $G(p^*(j)) \cdot r = Gp^*(j) \cdot G(m) \cdot \eta(\mathrm{B}) = G(p^*(j) \cdot m) \cdot \eta(\mathrm{B}) = G(q^*(j)) \cdot \eta(\mathrm{B}) = G(\nu(D(j)) \cdot Fq(j)) \cdot \eta(\mathrm{B}) = G\nu(D(j)) \cdot GFq(j) \cdot \eta(\mathrm{B}) = G\nu(D(j)) \cdot \eta(GD(j)) \cdot q(j) = 1_{GD(j)} \cdot q(j) = q(j)$ for each j of $O(J)$. Thus we have a morphism $r: \mathrm{B} \to G(\mathrm{A}^*)$ such that $Gp^*(j) \cdot r = q(j)$ for each j of $O(J)$. To show that this r is unique, let $s: \mathrm{B} \to G(\mathrm{A}^*)$ also satisfy $Gp^*(j) \cdot s = q(j)$ for each j. Then $q^*(j) = \nu(D(j)) \cdot Fq(j) = \nu(D(j)) \cdot F(Gp^*(j) \cdot s) = \nu(D(j)) \cdot FGp^*(j) \cdot F(s) = p^*(j) \cdot \nu(\mathrm{A}^*) \cdot F(s)$ for each j. The uniqueness statement for m shows then that $m = \nu(\mathrm{A}^*) \cdot F(s)$. So $r = G(m) \cdot \eta(\mathrm{B}) = G\nu(\mathrm{A}^*) \cdot GF(s) \cdot \eta(\mathrm{B}) = G\nu(\mathrm{A}^*) \cdot \eta(G(\mathrm{A}^*)) \cdot s$

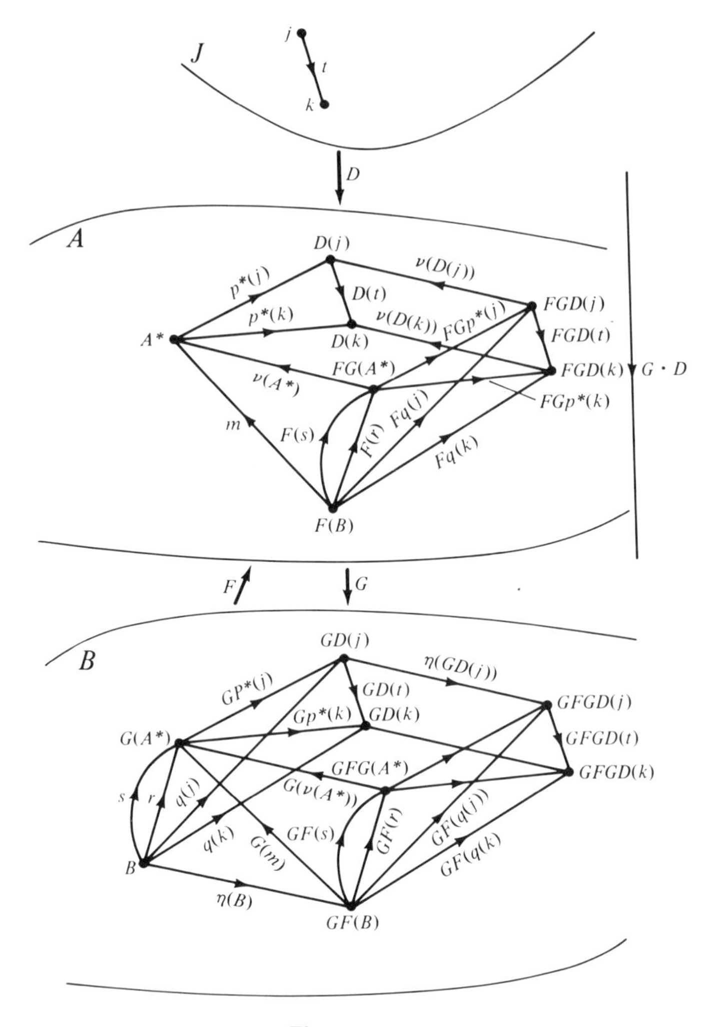

Figure 7.7

$= 1_{G(A^*)} \cdot s = s$. Thus the r is unique and we have proved that $(G(A^*), Gp^*)$ is a limit of GD in B.

(b) Let A be a complete category, G: $A \to B$ a functor that preserves all limits in A. If a certain object B of B satisfies the ssc relative to G, we shall construct a shortest G-path at B. (See Figure 7.8.)

First we choose a set of G-paths at B $[(h(k), A(k)]$, k in K, as given by the ssc. Since A is complete, the set of objects $[A(k)]$ has a product in A; that is, there is a limit $(A(o), p)$ for the functor $A\#$: $K \to A$ where K is considered a small discrete category and $A\#(k)$ is taken as $A(k)$ for each k of $O(K)$. Since G preserves limits, $(GA(o)), Gp)$ must be a limit

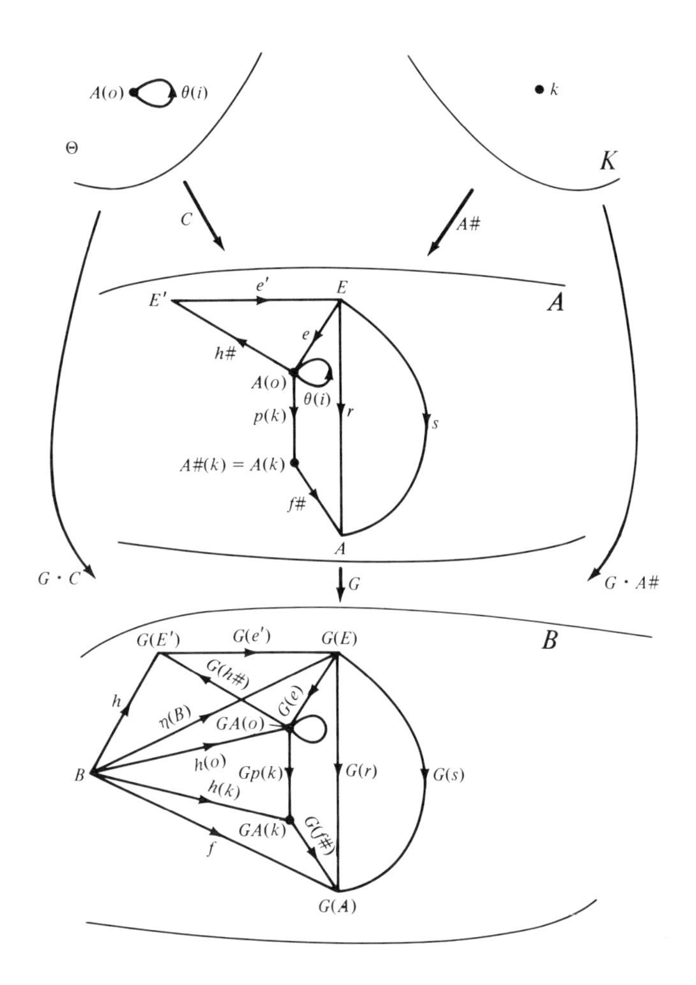

Figure 7.8

of $GA\#$ in B. Evidently (B, h) is a predecessor of this functor $GA\#$; hence there is a unique morphism $h(o)$: B $\rightarrow$ GA(o) [$=GA\#(o)$] such that

$$Gp(k) \cdot h(o) = h(k). \qquad (7.5.1)$$

There is a set [$\theta(i)$: i in I] of morphisms from A(o) to itself in A for which $G(\theta(i)) \cdot h(o) = h(o)$. This set includes $1_{A(o)}$ and is closed for compositions; hence we can consider these as the morphisms and A(o) as the single object of a subcategory Θ of A. The inclusion functor C: $\Theta \rightarrow A$ must have a limit (E, e) in A, with e: E $\rightarrow$ A(o); e is usually called a multiple-equalizer for the morphisms [$\theta(i)$: i in I]; it can be seen to be a monomorphism in A. Also, since G preserves limits, $(G(\mathrm{E}), G(e))$ would be a limit of $G \cdot C$ or a multiple-equalizer of $(G(\theta(i))$: i in $I)$. Since $G(\theta(i)) \cdot h(o) = h(o) = G(\theta(j)) \cdot h(o)$ for each pair i, j from I, it follows that $(\mathrm{B}, h(o))$ is a predecessor of $G \cdot C$ in B. Hence there must be a unique morphism $\eta(\mathrm{B})$: $\mathrm{B} \to G(\mathrm{E})$ such that

$$G(e) \cdot \eta(\mathrm{B}) = h(o). \tag{7.5.2}$$

We are going to show that $(\eta(\mathrm{B}), \mathrm{E})$ is a shortest G-path at B. It is surely a G-path at B. Let (f, A) be any G-path at B. From the choice of K we have a k in K and an $f\#$: $\mathrm{A}(k) \to \mathrm{A}$ such that

$$G(f\#) \cdot h(k) = f. \tag{7.5.3}$$

Using the results (7.5.1), (7.5.2), and (7.5.3) together, we get $f = G(f\#) \cdot h(k) = G(f\#) \cdot G(p(k)) \cdot h(o) = G(f\#) \cdot G(p(k)) \cdot G(e) \cdot \eta(\mathrm{B}) = G(f\# \cdot p(k) \cdot e) \cdot \eta(\mathrm{B})$. That is, we have a morphism $r = f\# \cdot p(k) \cdot e$ from E to A such that $f = G(r) \cdot \eta(\mathrm{B})$; to show that this r is unique, suppose that s: $\mathrm{E} \to \mathrm{A}$ also satisfies $f = G(s) \cdot \eta(\mathrm{B})$. The pair of morphisms r, s in A both from E to A must have an equalizer (E', e') since A is complete; and because G preserves limits in A, $(G(\mathrm{E}'), G(e'))$ must be an equalizer for $(G(r), G(s))$ in B. By assumption $G(r) \cdot \eta(\mathrm{B}) = G(s) \cdot \eta(\mathrm{B})$; hence, there must be a unique morphism h: $\mathrm{B} \to G(\mathrm{E}')$ such that

$$\eta(\mathrm{B}) = G(e') \cdot h. \tag{7.5.4}$$

From the choice of the family K and the definition of the product $\mathrm{A}(o)$ of the family $[\mathrm{A}(k)]$ it is clear that the one-element family $[(h(o), \mathrm{A}(o))]$ must also satisfy the ssc for B relative to G. Hence for the G-path (h, E') at B, there must be a morphism $h\#$: $\mathrm{A}(o) \to \mathrm{E}'$ such that

$$h = G(h\#) \cdot h(o). \tag{7.5.5}$$

Using in order (7.5.5), (7.5.4), and (7.5.2), we get $G(e \cdot e' \cdot h\#) \cdot h(o) = G(e) \cdot G(e') \cdot G(h\#) \cdot h(o) = G(e) \cdot G(e') \cdot h = G(e) \cdot \eta(\mathrm{B}) = h(o)$. Thus $(e \cdot e' \cdot h\#)$ must belong to the family Θ; since $1_{A(o)}$ also belongs to the family, and (E, e) is a multiple-equalizer for the family, we have $(e \cdot e' \cdot h\#) \cdot e = 1_{\mathrm{A}(o)} \cdot e = e = e \cdot 1_{\mathrm{E}}$. Using the fact that e is a monomorphism, hence left-cancelable, we deduce that $e' \cdot h\# \cdot e = 1_{\mathrm{E}}$; thus e' is a retraction and a regular monomorphism (being an equalizer); this implies that e' is an isomorphism. But then (E', e') is an equalizer of (r, s) means that $r = s$. Thus r is unique, and we have a shortest G-path $(\eta(\mathrm{B}), \mathrm{E})$ at B. □

From this lemma we can deduce easily P. Freyd's adjoint functor theorem:

Theorem 7.5.1 *Given a complete category A, a functor G: $A \to B$ has a* *left-adjoint iff (a) G preserves all limits in A, and (b) each object* B *of B satisfies the ssc relative to G.*

PROOF. If G: $A \to B$ has a left-adjoint F: $B \to A$, part (a) of Lemma 7.5.1 shows that G preserves all limits in A; further, for any B of $O(B)$ we know, by Theorem 7.2.1, that there is a shortest G-path (f^*, A*). This one-element set [(f^*, A*)] is clearly enough to provide the solution set condition for B relative to G. Thus conditions (a) and (b) are necessary.

Conversely, assuming these conditions, by part (b) of Lemma 7.5.1 we are assured of a shortest G-path for each object of B, and that assurance, by Theorem 7.2.1, implies that G has a left-adjoint. Thus the conditions are also sufficient. □

We also state another form of the theorem:

Theorem 7.5.2 (The Special Adjoint Functor Theorem) *Given a functor G: A → B, G would have a left-adjoint provided that A is a complete, well-powered category with a coseparator and G preserves limits.*

7.6 EXAMPLES OF ADJOINTS

7.6.1. The Algebras

Let F^* be a family of (finitary) operators, A some (nonnull) set, and D^* a set of pairs of elements from the free F^*-algebra $P(F^*, A)$. Let F, D be subsets, respectively, of F^*, D^*. Then F^*D^*-algebras are also FD-algebras and they are all set based. So we have the following (inclusion or forgetful) functors; C^*: $F^*D^*A \to F^*A$, C: $FDA \to FA$, G^*: $F^*D^*A \to FDA$, G: $F^*A \to FA$, H^*: $F^*A \to S$, and H: $FA \to$ S.

Lemma 7.6.1 *All the functors* C^*, C, G^*, G, H^*, H *and the composites* $H^* \cdot C^*$, $H \cdot C$, $G \cdot C^* = C \cdot G^*$, *and* $H \cdot G \cdot C^* = H^* \cdot C^*$ *have left-adjoints.*

PROOF. We already know from Lemmas 1.3.2 and 1.3.3 that each object of S has a shortest H-path (to FA) and a shortest $H \cdot C$-path (to FDA). Therefore H and $H \cdot C$, and similarly H^* and $H^* \cdot C^*$, have left-adjoints. To get one for C, let (X, F) be an arbitrary object of FA; for each set map g: $A \to X$ there is an associated F-homomorphism g^* of $(P(F, A), F)$ in (X, F). For each pair (p, q) from D ($\subseteq P(F, A) \times P(F, A)$) and each set map g: $A \to X$ we get a pair of elements $(g^*(p), g^*(q))$ in X; let E be the smallest F-congruence on (X, F) that contains all these pairs, and let p be the canonical F-homomorphism of the F-algebra (X, F) on the FD-algebra $(X/E, F)$. Then it is not hard to see that $(p, (X/E, F))$ is a shortest C-path at (X, F)). So C has a left-adjoint, and similarly, C^* has a left-adjoint.

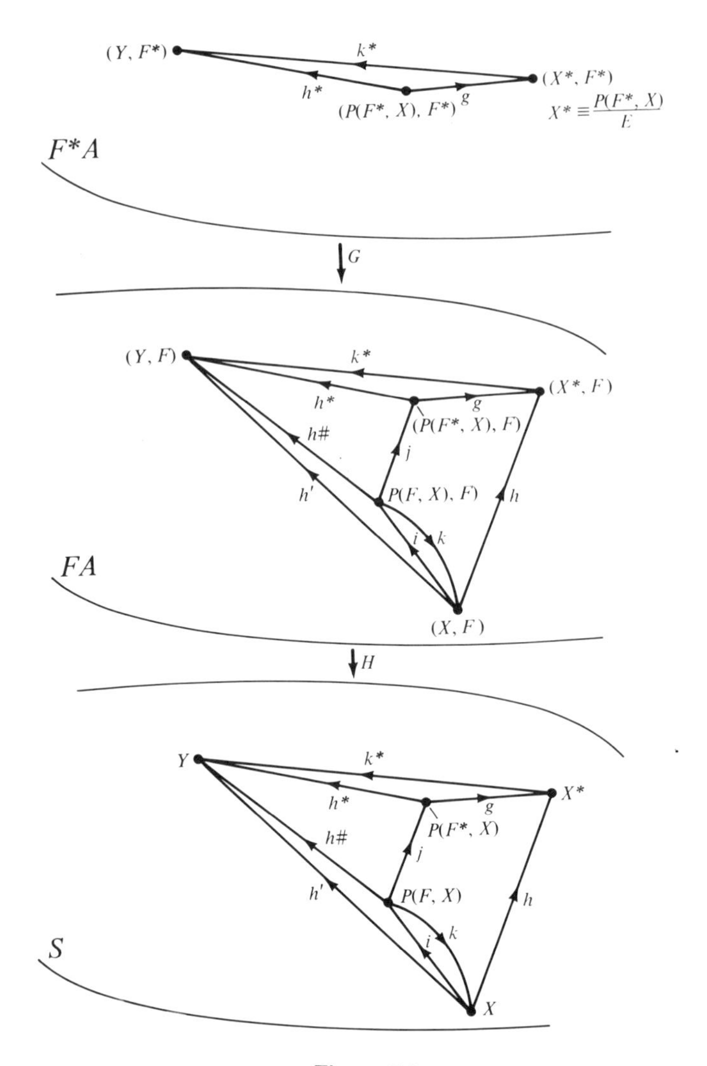

Figure 7.9

We next look at G: $F^*A \to FA$. (See Figure 7.9.) Let (X, F) be any object from FA. Form first the free F^*-algebra $(P(F^*, X), F^*)$. Surely $P(F^*, X)$ contains $P(F, X)$, since F^* contains F. Also, there is an F-homomorphism k: $(P(F, X), F) \to (X, F)$ with $k[(x)] = x$. The associated F-congruence $\theta = (k \cdot k^r)$ is also a relation on the set $P(F^*, X)$. Let E be the smallest F^*-congruence on the F^*-algebra $(P(F^*, X), F^*)$ containing this relation θ, and let g be the associated F^*-homomorphism of $(P(F^*, X), F^*)$ on the quotient algebra $(P(F^*, X)/E, F^*)$. We now describe an F-homomorphism h: $(X, F) \to (X^*, F^*) = (P(F^*, X)/E, F^*)$. First we consider the composition $g \cdot j$ of the F-homomorphism j (the inclusion): $P(F, X) \to P(F^*, X)$ followed by the map g considered as an F-homomorphism of $(P(F^*, X), F)$ in (X^*, F). For a pair of elements p, q from $P(F, X)$, if $k(p) = k(q)$, then (p, q) is in θ; hence $(j(p), j(q))$ is in E, by the definition of E, and $(g \cdot j)(p) = (g \cdot j)(q)$. Since k is a surjective homomorphism, there is a unique morphism h: $(X, F) \to (X^*, F)$ such that $h \cdot k = g \cdot j$. We show that $(h, (X^*, F^*))$ is a shortest G-path at (X, F). For if $(h', (Y, F^*))$ is another G-path at (X, F), so h': $(X, F) \to (Y, F)$ in FA, if i: $x \to [(x)]$ is the injective map of X in $P(F, X)$, the definition of $P(F^*, X)$ implies that there is a unique h^*: $(P(F^*, X), F^*) \to (Y, F^*)$ such that $h' = h^* \cdot j \cdot i$. Since $k \cdot i = 1_X$, we have $h' = (h' \cdot k) \cdot i$ also. Thus both $(h^* \cdot j)$ and $(h' \cdot k)$ are F-homomorphisms from $(P(F, X), F)$ to (Y, F), such that $h' = (h^* \cdot j) \cdot i$ and $h' = (h' \cdot k) \cdot i$. But the definition of the free F-algebra $(P(F, X), F)$ would ensure that there is a unique morphism $h\#$: $(P(F, X), F) \to (Y, F)$ for which $h' = h\# \cdot i$. Hence $h^* \cdot j = h\# = h' \cdot k$. It now follows that if (p, q) is in θ, $h^*j(p) = h'k(p) = h'k(q) = h^*j(q)$, since $k(p) = k(q)$. This means that the congruence for F^* determined by the homomorphism h^* contains the pair $(j(p), j(q))$ from $PF^*(X)$ whenever (p, q) is in θ; hence this congruence must contain E, which is the smallest congruence for F^* containing θ. Thus there is a homomorphism k^* from $(X^*, F^*) = (P(F^*, X)/E, F^*)$ to (Y, F^*) such that $k^* \cdot g = h^*$; hence $k^* \cdot h = k^* \cdot g \cdot j \cdot i = h^* \cdot j \cdot i = h'$. The uniqueness of the k^* for which $k^* \cdot h = h'$ follows from the fact that $k' \cdot h = h'$, for some k': $(X^*, F^*) \to (Y, F^*)$ would imply that $k' \cdot g \cdot j \cdot i = k' \cdot h = h' = h^* \cdot j \cdot i$. Because the morphism $k' \cdot g$ is also from $(P(F^*, X), F^*)$ to (Y, F^*), the uniqueness assertion about h^* implies that $k' \cdot g = h^*$. Hence $k' \cdot g = k^* \cdot g$; but g is a surjective map and so an epimorphism. Hence it is right-cancelable, giving $k' = k^*$. Thus we have proved that $(h, (X^*, F^*))$ is a shortest G-path at (X, F). So G has a left-adjoint.

Since C^* and G both have left-adjoints, by Theorem 7.4.1, there is a left-adjoint for $G \cdot C^* = C \cdot G^*$ also.

For the functor G^*: $F^*D^*A \to FDA$ we get the left-adjoint by using the last construction to obtain, for each FD-algebra (X, F) a shortest G-path, considering it an F-algebra. Then this too would be a shortest G^*-path, as can easily be seen. Hence each object of FDA has a shortest G^*-path, and G^* has a left-adjoint. □

While Lemma 7.6.1 gives a shortest G^*-path for each FDA-algebra, leading to an F^*D^*A-algebra, there is an associated "immersion problem" that is not generally solvable, namely, what FDA-algebra can be embedded isomorphically in an F^*D^*A-algebra? In the simple case, for instance, of the semigroups and the groups, not all semigroups can be viewed as subsemigroups of groups (up to an isomorphism, of course). The needed conditions can generally be extracted by using the following general result:

Lemma 7.6.2 *If F, D are subfamilies of F^*, D^*, respectively, where F^* is a family of (algebraic) operations and D^* a set of pairs from $P(F^*, A)$ for some set A and G^*: $F^*D^*A \to FDA$ is the forgetful functor, then an object (X, F) of FDA can be isomorphic with an FD-subalgebra of an object (Y, F^*) of F^*D^*A iff there is a shortest G^*-path $(h, (X^*, F^*))$ for (X, F) with h a monomorphism.*

PROOF. If $(h, (X^*, F^*))$ is a shortest G^*-path at (X, F) with h a monomorphism, then h must be an injective set map; then it would also be an isomorphism of (X, F) with an FD-subalgebra of (X^*, F^*). Thus the condition is sufficient.

Conversely, if there is an isomorphism j of (X, F) with an F-subalgebra of an F^*D^*-algebra (Y, F^*), we have a G^*-path $(j, (Y, F^*))$ at (X, F); also, Lemma 7.6.1 assures that there is a shortest G^*-path $(h, (X^*, F^*))$ at (X, F). There must then be a unique θ: $(X^*, F^*) \to (Y, F^*)$ in F^*D^*A such that $G^*(\theta) \cdot h = j$. Then, from the fact that j is a monomorphism it follows that h also is one (for $h \cdot r = h \cdot s$ implies $j \cdot r = G^*(\theta) \cdot h \cdot r = G^*(\theta) \cdot h \cdot s = j \cdot s$, from which $r = s$ follows). □

In the exercises at the end of this chapter, Lemma 7.6.2 is used in discussing the Malčev conditions for a semigroup to be embeddable in a group.

7.6.2. Preorder, Bitopology, and Semiuniformity

A semiuniformity U for a set X determines a reverse semiuniformity U^r; each of these defines a topology $T(U)$ or $T(U^r)$ on X, and U determines a preorder on X, namely, $\leq(U) = [(x, y): (y, x) \in \text{each } U(j)]$; the preorder $\leq(U^r)$ that U^r determines on X is clearly the reverse of $\leq(U)$. These preorders on X can also be directly related to the topologies $T(U)$ or $T(U^r)$. In fact, given a topological space (X, T), we define the preorder $\leq(T)$ on X by $x \leq (T)y$ iff [$\forall$G in T: y in G implies x in G]; then we denote the associated equivalence $\leq(T) \cap \geq(T)$ by $\equiv (T)$, calling it topological equivalence in (X, T). It is the same as equality iff the space is T_0. Now $\leq(U)$ is the same as $\leq(T(U))$; so the two topologies $T(U)$ and $T(U^r)$ on X determine a preorder and its reverse on X. We now introduce the notion of a *bitopological space* (X, T, T'); it is a set X with two topologies T, T' on it such that $\leq(T)$ and $\leq(T')$ are reverse preorders on X. A map f: $X \to Y$ is a morphism between two bitopological spaces (X, T, T') and (Y, S, S') if f is a continuous map of (X, T) in (Y, S) and of (X, T') in

(Y, S'). These then give rise to a category BT of bitopological spaces and their morphisms. There is an obvious functor $B: SU \to BT$; for an (X, U) in $O(SU)$, $B(X, U) = (X, T(U), T(U^r))$, and B takes a morphism f in SU to a morphism f in BT (based on the same set map f). From BT we have two functors to T: one, L, takes (X, T, T') to (X, T), and the other, R, takes (X, T, T') to (X, T'). There are also two functors $\mathring{\leq}: T \to PO$ and $\mathring{\geq}: T \to PO$, which take a typical (X, T) from $O(T)$ to $(X, \leq(T))$ and to $(X, \geq(T))$, respectively. These functors L, R, $\mathring{\leq}$, and $\mathring{\geq}$ also take morphisms to morphisms with the same set-based mapping relations. We also use in the next lemma, the inclusion functor C from CSU to SU where CSU denotes the full subcategory of SU with objects that are T_0, complete semiuniform spaces.

Lemma 7.6.3 *Each of the following functors and the composites of these, when defined, have left-adjoints;* $\mathring{\leq}: T \to PO$, $\mathring{\geq}: T \to PO$, $L: BT \to T$, $R: BT \to T$, $B: SU \to BT$, *and* $C: CSU \to SU$. *Of the six listed,* C *is a retractor, while the first four are coretractors.*

PROOF. Once we prove that left-adjoints exist for the six listed functors, it would follow, by our composition theorem, that composites of these when defined also have left-adjoints. For each of the typical functors F listed we prove that a left-adjoint exists by showing that each object of the codomain of F has a shortest F-path.

The Functors $\mathring{\leq}$ *and* $\mathring{\geq}$. Given a preordered set $(X, \leq)$, we associate to it the space $(X, H) = \wedge(X, \leq)$, with H equal to [the family of initial sets in $(X, \leq)$]. Since this family H is closed for arbitrary unions and intersections and contains $\varnothing$ and X, it does give a topology H for X. Also, the preorder that H defines is the same as $\leq$: for $x \leq y$ implies that each set of H that contains y contains x, so that $x\ (\leq(H))\ y$; whereas $x\ (\leq(H))\ y$ implies that each set of H containing y must also contain x. Since $\wedge(y) = [z: z \leq y]$ belongs to H and contains y, it must contain x; that is, $x \leq y$. We claim that $(1_X, (X, H))$ is a shortest $\mathring{\leq}$-path at $(X, \leq)$: by the last remark it is a $\mathring{\leq}$-path all right. If $g: (X, \leq) \to \mathring{\leq}(Y, G)$ is another $\mathring{\leq}$-path, g is a monotone map from $(X, \leq)$ to $(Y, (\leq(G)))$; hence, for each B of G, $g^r(B)$ must be an initial set of $(X, \leq)$, since y in $g^r(B)$ and $x \leq y$ imply $g(x)\ (\leq(G))\ g(y)$ and $g(y)$ is in B, but then, from the definition of $\leq(G)$, $g(x)$ must be in B or x must be in $g^r(B)$. Hence $g: (X, H) \to (Y, G)$ is a continuous map with the property that $\mathring{\leq}(g) \cdot 1_X = g$. The uniqueness of such a map from (X, H) to (Y, G) follows from the fact that the other factor is 1_X in the factorization of g. These prove that $(1_X, (X, H))$ is a shortest $\mathring{\leq}$-path at $(X, \leq)$, and that $\mathring{\leq}$ is a coretractor, since $\eta(X, \leq)$ is now $1_{(X, \leq)}$.

The proof of the case of $\mathring{\geq}$ is similar; we use the topology H^* for X defined from $(X, \leq)$ by taking all final subsets of $(X, \leq)$ as sets of H^* This gives the functor $\vee$, which is a left-adjoint of $\mathring{\geq}$.

The Functors L and R. Starting from a space (X, T), we associate to X a second topology T' consisting of all the final subsets of $(X, (\leq(T)))$, [so that $(X, T') = \bigvee(\mathring{\leq}(X, T))$]. Then the preorder on X defined by T' is the reverse of that defined by T, so that (X, T, T') is a bitopological space. We claim that $(1_X, (X, T, T'))$ is a shortest L-path at (X, T); it is clearly an L-path. For any L-path $(g, (Y, S, S'))$ at (X, T), g being continuous from (X, T) to (Y, S), we check that g is also continuous from (X, T') to (Y, S'), so that g: $(X, T, T') \to (Y, S, S')$ is a morphism in BT. From the definition of $(\leq(S')) = (\leq(S))^r$ we know that any set B' of S' is an initial set of $(Y, (\leq(S')))$ or a final subset of $(Y, (\leq(S)))$. But then the continuity of g from (X, T) to (Y, S) implies that g: $(X, (\leq(T))) \to (Y, (\leq(S)))$ is a monotone map, so that g^r [a final subset of $(Y, (\leq(S)))$] is equal to a final subset of $(X, (\leq(T)))$, that is, belongs to T'. So we have g continuous from (X, T') to (Y, S'). This g from (X, T, T') to (Y, S, S') in BT is surely the unique morphism g^* with the property that $L(g^*) \cdot 1_X = g$. These prove that $(1_x, (X, T, T'))$ is the shortest L-path at (X, T); thus L has a left-adjoint. Here also $\eta(X, T)$ is $1_{(X,T)}$, so that L is a coretractor.

The functor R is treated similarly.

The Functor B. For an (X, T, T') from BT, we consider the family of the semiuniformities [U_j: $T(U_j)$ is coarser than T, and $T(U_j{}^r)$ is coarser than T']; the family is nonnull because the coarsest semiuniformity [$(X \times X)$] on X is in it. If U^* is the lattice product of this family (under the preorder of being "finer than"), it is not hard to show that $(1_X, (X, U^*))$ gives a shortest B-path at (X, T, T'). So B has a left-adjoint. This B is further a coretractor.

The Functor C. Theorem 1.6.1 proves in effect that for any (X, U) of SU there is a shortest C-path $(h, (X\#, U\#))$; and if (X, U) were a T_0 complete space, it could be verified that h is an isomorphism in SU; so C is a retractor. □

We already defined topological equivalence $\equiv(T)$ in any space (X, T) by $x \equiv(T)\ y$ iff a set from T contains either of x, y only when it contains both of them. The passage from X to $X/\equiv(T)$ gives rise to a space $(X/\equiv(T), T/\equiv(T))$ where the topology also is obtained by taking the images of the sets of T by the canonical surjective map of X on $X/\equiv(T)$. The passage by the canonical surjective continuous map p: $(X, T) \to (X/\equiv, T/\equiv)$ is called *identification of topologically equivalent points* in (X, T). This leads to some adjoints, as we see next.

Lemma 7.6.4 *Denoting by T_1 and T_2 the categories $(T_1' \cap T_0)$ and $(T_2' \cap T_0)$, respectively, and by (CT_2) the category $(C \cap T_2)$, we have the following inclusion functors, which are all retractors and have left-adjoints: C_0: $T_0 \to T$, C_1: $T_1 \to T_1'$, C_2: $T_2 \to T_2'$, and C^*: $(CT_2) \to T_2$.*

PROOF. When (X, T) is an object from T, T_1', or T_2', and $\equiv$ is topological equivalence defined on (X, T), and p the canonical surjection of (X, T) on the quotient space by $\equiv$, it is not hard to see that $(p, (X/\equiv, T/\equiv)$ is a shortest C_0-, C_1-, or C_2-path in the three cases; and since p becomes a homeomorphism when (X, T) is a T_0-space, all three functors are retractors.

To get a left-adjoint for the inclusion functor C^*: $(CT_2) \to T_2$, we shall utilize the *special adjoint functor theorem*. Note first that for any family of T_2-spaces $[(X_j, \mathbf{G}_j)$: j in $J]$, the product space $(X, \mathbf{G})$ is compact when each of the $(X_j, \mathbf{G}_j)$ is compact, and for a pair of parallel morphisms (or continuous maps) f, g both from a T_2-space $(X, \mathbf{G})$ to a T_2-space $(Y, \mathbf{H})$, an equalizer is the inclusion map in $(X, \mathbf{G})$ of the subspace Z of $(X, \mathbf{G})$ consisting of the z of X for which $f(z) = g(z)$. And since this subspace is closed in $(X, \mathbf{G})$, it would be compact if $(X, \mathbf{G})$ were so.

These remarks show that (CT_2) is a complete category (by Theorem 6.4.1(b)); and they further imply that C^* preserves all limits (by Exercise 6.2(ii)). As (CT_2) has a coseparator, namely, the space $I = [0, 1]$ (which is a consequence of the complete regularity of a compact Hausdorff space), the special adjoint functor theorem applies here. So C^* has a left-adjoint. Since C^* is an inclusion functor, it is not hard to see that C^* is a retractor (even a reflector).

The left-associate of C^* is a functor M, which associates to each T_2-space a compact T_2-space known as the *Stone–Cěch compactification* of the original T_2-space. If $(X, \mathbf{G})$ is the T_2-space, the compactification can be shown to be the closure of a subspace $X\#$ of a power space I^K, where K denotes the family of all continuous functions from $(X, \mathbf{G})$ to I, and $X\#$ consists of the points of the form $x\#$ for x in X, when we set $x\#(f) = f(x)$ for each f from K. □

Further examples of adjoints involving algebras over a semiuniform space, over a bitopological space, and over a preordered set are treated in Appendix 1. This requires a proper definition of co- and contravariance of algebraic operations relative to the other structure. Interested readers are referred to Appendix 1.

7.7 MONADS

Starting with a near-equivalence $(\eta, \nu; G, F)$ for the pair of categories (A, B), we can derive the following: a functor $H = G \cdot F$: $B \to B$, the natural transformation η: $1_B \to H$, and the natural transformation $\delta = 1_G \# \nu \# 1_F$: $H \cdot H \to (G \cdot 1_A \cdot F = G \cdot F =)H$; further, from the properties of the near-equivalence

1. for each B of $O(B)$, $\nu F(\mathrm{B}) \cdot F\eta(\mathrm{B}) = 1_{F(\mathrm{B})}$,
2. for each A of $O(A)$, $G\nu(\mathrm{A}) \cdot \eta G(\mathrm{A}) = 1_{G(\mathrm{A})}$,

we get the following:

1′. for each B of $O(B)$, $G\nu F(\mathrm{B}) \cdot GF\eta(\mathrm{B}) = G1_{F(\mathrm{B})} = 1_{GF(\mathrm{B})}$,
2′. for each B of $O(B)$, $G\nu F(\mathrm{B}) \cdot \eta GF(\mathrm{B}) = 1_{GF(\mathrm{B})}$,

which in turn give the commutativity of the following diagram of natural transformations:

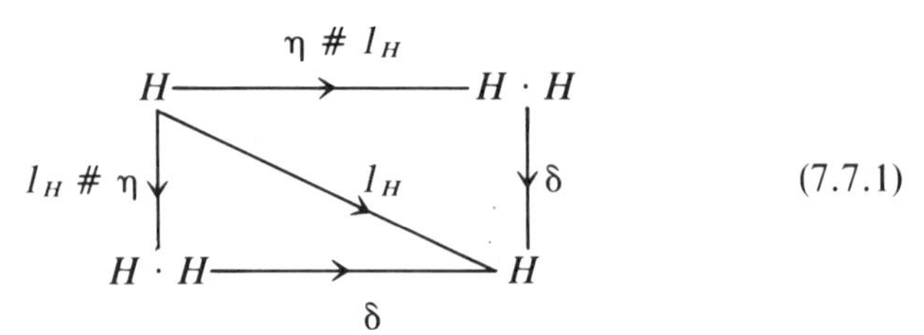

(7.7.1)

Also, since ν is a natural transformation from $F \cdot G$ to 1_A, for any B of $O(B)$ we have the commutative diagram:

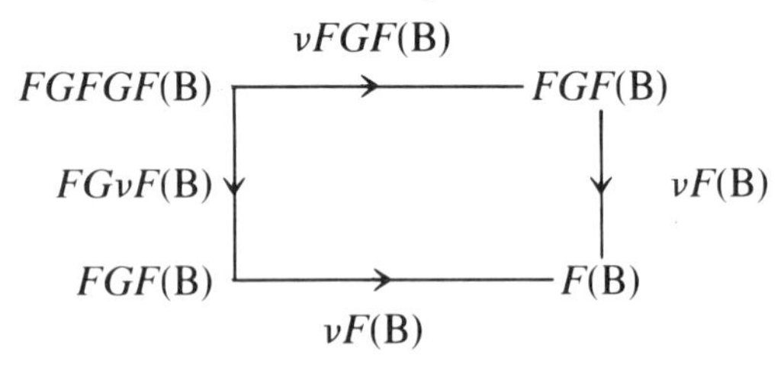

The transform of this commutative diagram by the functor G gives rise to a commutative diagram that, in turn, being true for all B of $O(B)$, implies the commutativity of the following diagram of natural transformations:

$$\begin{array}{ccc} H \cdot H \cdot H & \xrightarrow{1_H \# \delta} & H \cdot H \\ {\scriptstyle \delta \# 1_H} \downarrow & & \downarrow {\scriptstyle \delta} \\ H \cdot H & \xrightarrow{\delta} & H \end{array} \qquad (7.7.2)$$

The properties that we proved in the foregoing for H, η, and δ motivate the following:

Definition 7.7.1 Given a category B, we call (H, η, δ) a *monad* (some call it a *triple*) on B if $H: B \to B$ is a functor, and $\eta: 1_B \to H$ and $\delta: H \cdot H \to H$ are natural transformations such that diagrams (7.7.1) and (7.7.2) of natural transformations are both commutative.

Starting with the category B, if we have a pair of functors $G: A \to B$ $F: B \to A$ with F a left-adjoint of G, then we know that this gives rise to a near-equivalence $(\eta, \nu; G, F)$ for (A, B); hence, as above, there is an associated monad (H, η, δ) on B. We shall say that this monad is defined by (A, F, G) when it is so derived from these three. The main result is that any monad on B can be so derived from a *triad* of the form (A, F, G), often in more than one way. To compare these triads that give rise to the same monad (H, ν, δ), we shall define for these a preorder $\leq$ by setting $(A, F, G) \leq (A', F', G')$ iff there is a functor $K: A \to A'$ such that $K \cdot F = F'$, $G' \cdot K = G$. With this relation for these triads we can now state the main result:

Theorem 7.7.1 *Given a monad (H, η, δ) on B, there exist triads (A, F, G) that define this monad on B; and under the preorder among these triads defined earlier, there exists a least triad $(A^\circ, F^\circ, G^\circ)$ and a greatest triad (A^*, F^*, G^*).*

PROOF. We start by constructing the triad $(A^\circ, F^\circ, G^\circ)$ first; $O(A^\circ) = O(B)$. For A, B from $O(A^\circ) = O(B)$, $\hom_{A^\circ}(\mathrm{A}, \mathrm{B}) = [f$ in $\hom_B(H(\mathrm{A}), H(\mathrm{B}))$ for which $f \cdot \delta(\mathrm{A}) = \delta(\mathrm{B}) \cdot H(f)]$; and the composition of morphisms in A° is the same as their composition in B. It is simple enough to check that this gives indeed a category A°. We define $F^\circ: B \to A^\circ$ by $F^\circ(\mathrm{B}) = \mathrm{B}$ for each object B of B, $F^\circ(g) = H(g)$ for a g: A $\to$ B in $M(B)$. We define G°: $A^\circ \to B$ by $G^\circ(\mathrm{A}) = H(\mathrm{A})$ for an A of $O(A^\circ) = O(B)$; $G^\circ(f) = f$ for an f: A $\to$ B of $M(A^\circ)$, so that $f \in \hom_B(H(\mathrm{A}), H(\mathrm{B}))$! We now check that $(\eta, \delta; G^\circ, F^\circ)$ is a near-equivalence for (A°, B) that determines the monad (H, η, δ) on B. Clearly $G^\circ \cdot F^\circ = H$; and by hypothesis $\eta: 1_B \to H = G^\circ \cdot F^\circ$ is a natural transformation. For any A from $O(A^\circ) = O(B)$, $F^\circ(G^\circ(\mathrm{A})) = F^\circ(H(\mathrm{A})) = H(\mathrm{A})$, and $\delta(\mathrm{A}): HH(\mathrm{A}) \to H(\mathrm{A})$ in $M(B)$ can be considered a morphism in A° from $H(\mathrm{A})$ to A; for this satisfies the required condition to belong to $\hom_{A^\circ}(H(\mathrm{A}), \mathrm{A})$, namely, $\delta(\mathrm{A}) \cdot \delta(H(\mathrm{A})) = \delta(\mathrm{A}) \cdot H\delta(\mathrm{A})$ because of the assumed commutativity of diagram (7.7.2). This makes $\delta: F^\circ \cdot G^\circ \to 1_{A^\circ}$ a natural transformation. Now to see that $(\eta, \delta; G^\circ, F^\circ)$ is a near-equivalence for (A°, B) we have to check that for each B of $O(B)$ we have $\delta F^\circ(\mathrm{B}) \cdot F^\circ\eta(\mathrm{B}) = 1_{F^\circ(\mathrm{B})}$ and for each A of $O(A^\circ)$ we also have $G^\circ\delta(\mathrm{A}) \cdot \eta G^\circ(\mathrm{A}) = 1_{G^\circ(\mathrm{A})}$; but these reduce to the relations $\delta(\mathrm{B}) \cdot H\eta(\mathrm{B}) = F^\circ(1_\mathrm{B}) = 1_{H(\mathrm{B})}$ and $\delta(\mathrm{A}) \cdot \eta H(\mathrm{A}) = 1_{H(\mathrm{A})}$, both of which follow from the commutativity of (7.7.1). It is not hard to see that this near-equivalence does define the monad (H, η, δ).

Suppose next that we have some triad (A', F', G'), also defining the same monad (H, ν, δ) through the near-equivalence $(\eta, \nu'; G', F')$ for (A', B). We have to show that $(A^\circ, F^\circ, G^\circ) \leq (A', F', G')$. First we define the functor $K: A^\circ \to A'$ by setting, for an object A of A° and a morphism f: A $\to$ B in A°, $K(\mathrm{A}) = F'(\mathrm{A})$ and $K(f) = \nu'(F'(\mathrm{B})) \cdot F'(f) \cdot F'(\eta(\mathrm{A}))$. Since f is in $\hom_B(H(\mathrm{A}), H(\mathrm{B}))$, $F'(f)$ is in $\hom_{A'}(F'H(\mathrm{A}), F'H(\mathrm{B}))$ or $\hom_{A'}(F'G'F'(\mathrm{A}), F'G'F'(\mathrm{B}))$, so that the morphism $K(f)$ belongs to

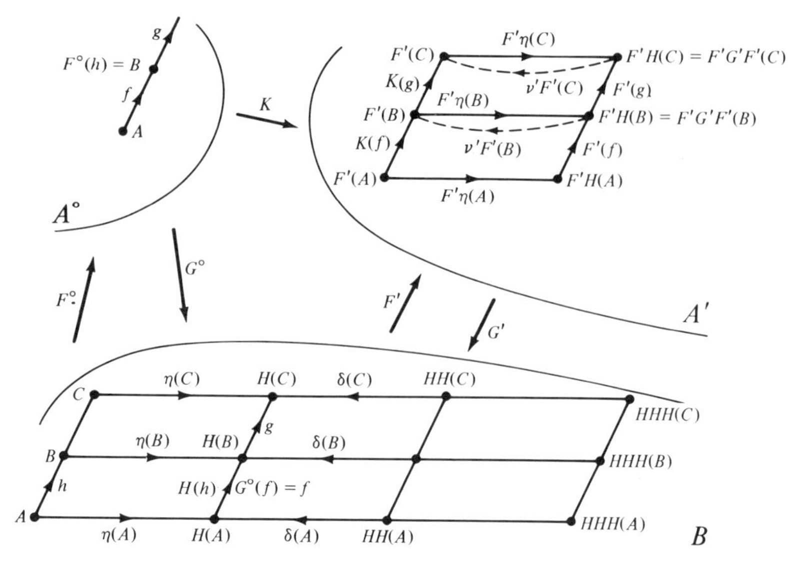

Figure 7.10

$\hom_{A'}(F'(\mathrm{A}), F'(\mathrm{B}))$ or $\hom_{A'}(K(\mathrm{A}), K(\mathrm{B}))$. To see that K is a functor, we check its effect on a 1_A of A° and on a composite $(g \cdot f)$ where $f\colon \mathrm{A} \to \mathrm{B}$ and $g\colon \mathrm{B} \to \mathrm{C}$ in A°. (See Figure 7.10.) $K(1_A) = \nu' F'(\mathrm{A}) \cdot F'(1_{H(\mathrm{A})}) \cdot F'(\eta(\mathrm{A})) = \nu'(F'(\mathrm{A}) \cdot F'(\eta(\mathrm{A})) = 1_{F'(\mathrm{A})} = 1_{K(\mathrm{A})} \cdot F'G'(K(g)) = F'G'[\nu'(F(C)) \cdot F'(g) \cdot F'(\eta(\mathrm{B}))] = F'[G'\nu'F'(C) \cdot G'F'(g)] \cdot F'H(\eta(\mathrm{B})) = F'[\delta(C) \cdot H(g)] \cdot F'H(\eta(\mathrm{B})) = F'(g \cdot \delta(\mathrm{B})) \cdot F'(H\eta(\mathrm{B})) = F'(g) \cdot F'[\delta(\mathrm{B}) \cdot H\eta(\mathrm{B})] = F'(g) \cdot F'(1_{H(\mathrm{B})}) = \mathrm{F}'(g)$. Hence we have $K(g) \cdot K(f) = K(g) \cdot \nu'F'(B) \cdot F'(f) \cdot F'\eta(\mathrm{A}) = \nu'F'(\mathrm{C}) \cdot F'G'K(g) \cdot F'(f) \cdot F'\eta(\mathrm{A})$, since $\nu'\colon F'G' \to 1_{A'}$ is a natural transformation, $= \nu'F'(\mathrm{C}) \cdot F'(g) \cdot F'(f) \cdot F'\eta(\mathrm{A}) = \nu'F'(\mathrm{C}) \cdot F'(g \cdot f) \cdot F'\eta(\mathrm{A}) = K(g \cdot f)$. Thus K is a functor from A° to A'.

We then have to check that $K \cdot F^\circ = F'$. Take a typical object A of B and morphism $g\colon \mathrm{A} \to \mathrm{B}$ in B; $K \cdot F^\circ(\mathrm{A}) = K(\mathrm{A}) = F'(\mathrm{A})$, and $K \cdot F^\circ(g) = K(H(g)) = \nu'F'(\mathrm{B}) \cdot F'H(g) \cdot F'\eta(\mathrm{A}) = \nu'F'(\mathrm{B}) \cdot F'[H(g) \cdot \eta(\mathrm{A})] = \nu'F'(\mathrm{B}) \cdot F'[\eta(\mathrm{B}) \cdot g]$, since $\eta\colon 1_B \to H$ is natural, $= \nu'F'(\mathrm{B}) \cdot F'\eta(\mathrm{B}) \cdot F'(g) = 1_{F'(\mathrm{B})} \cdot F'(g) = F'(g)$. Thus we do have $K \cdot F^\circ = F'$.

Then we check that $G' \cdot K = G^\circ$. Take again a typical object A of A° and typical morphism $f\colon \mathrm{A} \to \mathrm{B}$ of A°, so that f is in $\hom_B(H(\mathrm{A}), H(\mathrm{B}))$. Then $G' \cdot K(\mathrm{A}) = G' \cdot F'(\mathrm{A}) = H(\mathrm{A}) = G^\circ(\mathrm{A})$, while $G' \cdot K(f) = G'[\nu'F'(\mathrm{B}) \cdot F'(f) \cdot F'\eta(\mathrm{A})] = \delta(\mathrm{B}) \cdot G'F'(f) \cdot G'F'\eta(\mathrm{A}) = \delta(\mathrm{B}) \cdot H(f) \cdot H\eta(\mathrm{A}) = f \cdot \delta(\mathrm{A}) \cdot H\eta(\mathrm{A}) = f \cdot 1_{H(\mathrm{A})} = f$ in $B = G^\circ$ (f of A°).

We now go on to define the triad (A^*, F^*, G^*). The category A^* is defined by

$$O(A^*) = [(\mathrm{A}, a)\colon \mathrm{A} \text{ in } O(B),\ a\colon H(\mathrm{A}) \to \mathrm{A}, \text{ a morphism in } B \text{ such that } a \cdot \eta(\mathrm{A}) = 1_{\mathrm{A}} \text{ and } a \cdot H(a) = a \cdot \delta(\mathrm{A})];$$

$$\hom_{A^*}[(\mathrm{A}, a), (\mathrm{B}, b)] = [\text{morphisms } f\colon \mathrm{A} \to \mathrm{B} \text{ in } B \text{ such that } f \cdot a = b \cdot H(f)].$$

The morphisms in A^* compose as they do in B. With $1_{(\mathrm{A},a)}$ as 1_{A}, it is easy to see that A^* is indeed a category. We now define the functors F^*, G^*. For an object A of B and a morphism $f\colon \mathrm{A} \to \mathrm{B}$ in B we set $F^*(\mathrm{A}) = (H(\mathrm{A}), \delta(\mathrm{A}))$ and $F^*(f) = H(f)$. Since δ is natural, it follows that $F^*(f)\colon F^*(\mathrm{A}) \to F^*(\mathrm{B})$ in A^*. The fact that (H, η, δ) is a monad assures that $(H(\mathrm{A}), \delta(\mathrm{A}))$ is an object of A^*. Since H is a functor, it follows that this F^* is a functor from B to A^*. For an object (A, a) of A^* and a morphism $f\colon (\mathrm{A}, a) \to (\mathrm{B}, b)$ of A^*, we set $G^*(\mathrm{A}, a) = \mathrm{A}$ and $G^*(f) = f$ itself. Then G^* is a functor from A^* to B. This triad determines a near-equivalence $(\eta, \nu^*; G^*, F^*)$ for the pair (A^*, B) where $\nu^*(\mathrm{A}, a)\colon F^*G^*(\mathrm{A}, a) \to (\mathrm{A}, a)$ is taken to be $a\colon (H(\mathrm{A}), \delta(\mathrm{A})) \to (\mathrm{A}, a)$. Our assumption that $a \cdot \delta(\mathrm{A}) = a \cdot H(a)$ when (A, a) is in $O(A^*)$ ensures that this does give a morphism (a) in A^*. To see that this is a near-equivalence, for an object B of B, $\nu^*(F^*(\mathrm{B})) \cdot F^*(\eta(\mathrm{B})) = \nu^*(H(\mathrm{B}), \delta(\mathrm{B})) \cdot H(\eta(\mathrm{B})) \doteq \delta(\mathrm{B}) \cdot H(\eta(\mathrm{B})) = 1_{H(\mathrm{B})} = 1_{F^*(\mathrm{B})}$, while for any object (A, a) of A^*, $G^*(\nu^*(\mathrm{A}, a)) \cdot \eta(G^*(\mathrm{A}, a)) = G^*(a) \cdot \eta(\mathrm{A}) = a \cdot \eta(\mathrm{A}) = 1_{\mathrm{A}}$, since (A, a) is in $O(A^*)$, $= 1_{G^*(\mathrm{A},a)}$. So we have a near-equivalence, and the monad that it determines is (H, η, δ), since $G^*\nu^*F^*(\mathrm{B}) = G^*\nu^*(H(\mathrm{B}), \delta(\mathrm{B})) = G^*(\delta(\mathrm{B})) = \delta(\mathrm{B})$. Hence this triad (A^*, F^*, G^*) also determines (H, η, δ). (See Figure 7.11.)

If we now take any triad (A', F', G') determining this monad (H, η, δ) through a near-equivalence $(\eta, \nu'; F', G')$ for (A', B), we wish to prove that $(A', F', G') \leq (A^*, F^*, G^*)$. We have to get a functor $L\colon A' \to A^*$ such that $L \cdot F' = F^*$ and $G^* \cdot L = G'$. Since $\nu'\colon F'G' \to 1_{A'}$ is a natural equivalence, we have the morphism $\nu'(\mathrm{A}')\colon F'G'(\mathrm{A}') \to \mathrm{A}'$ for each object A' of A'; and then we show that $(G'(\mathrm{A}'), G'\nu'(\mathrm{A}'))$ is an object of A^*, and this we shall set as $L(\mathrm{A}')$. Now $G'\nu'(\mathrm{A}')\colon G'F'G'(\mathrm{A}') \to G'(\mathrm{A}')$, or $G'\nu'(\mathrm{A}')\colon H \cdot G'(\mathrm{A}') \to G'(\mathrm{A}')$; further, $G'\nu'(\mathrm{A}') \cdot \eta(G'(\mathrm{A}')) = 1_{G'(\mathrm{A}')}$, by the near-equivalence property. The other property to be proved to ensure that $(G'(\mathrm{A}'), G'\nu'(\mathrm{A}'))$ belongs to $O(A^*)$ is that $G'\nu'(\mathrm{A}') \cdot \delta(G'(\mathrm{A}')) = G'\nu'(\mathrm{A}') \cdot HG'\nu'(\mathrm{A}')$. Since the near-equivalence defines the monad given, the last relation can be written as $G'\nu'(\mathrm{A}') \cdot (G'\nu'F')(G'(\mathrm{A}')) = G'\nu'(\mathrm{A}') \cdot G'F'G'\nu'(\mathrm{A}')$; but this last result follows by applying the functor G' to the relation $\nu'(\mathrm{A}') \cdot \nu'F'G'(\mathrm{A}') = \nu'(\mathrm{A}') \cdot F'G'\nu'(\mathrm{A}')$, which is true in A', since ν' is a natural equivalence from $F'G'$ to $1_{A'}$. Thus we

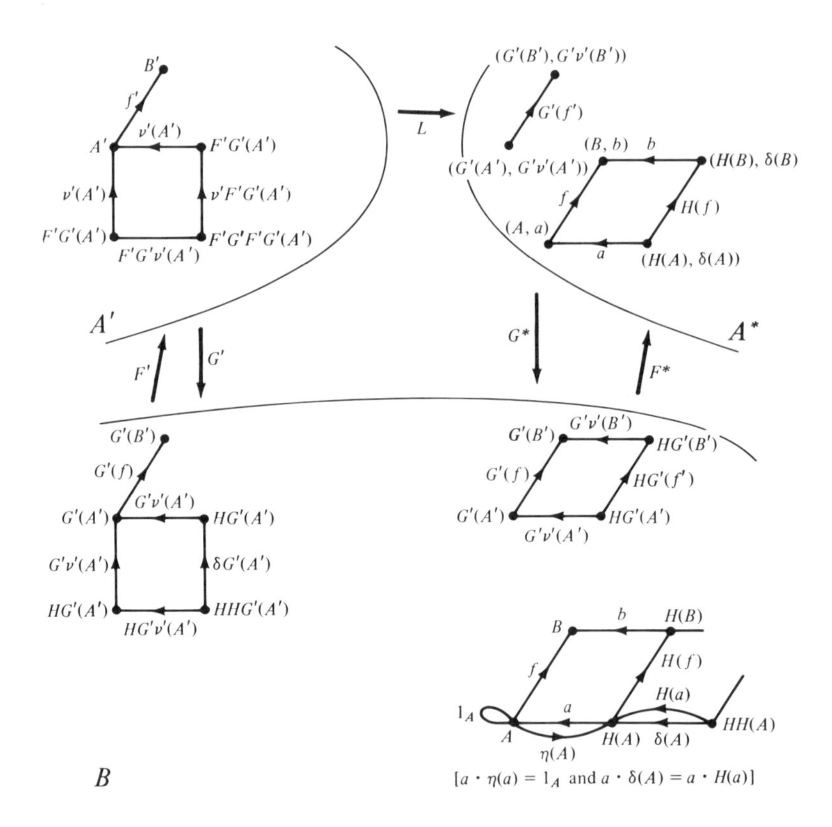

Figure 7.11

have now defined L for an object A′ of A'; for a morphism f': A′ → B′ of A' we set $L(f') = G'(f')$: $(G'(\mathrm{A}'), G'\nu'(\mathrm{A}')) \to (G'(\mathrm{B}'), G'\nu'(\mathrm{B}'))$. Since ν' is a natural transformation, it is seen that this $L(f')$ is a morphism in A^*. With these definitons we get a functor L: $A' \to A^*$.

We now check the relation $G^* \cdot L = G'$: for an object A′ of A' and a morphism f': A′ → B′ in A', $G^* \cdot L(\mathrm{A}') = G^*[(G'(\mathrm{A}'), G'\nu'(\mathrm{A}')] = G'(\mathrm{A}')$; and $G^* \cdot L(f') = G'(f')$. These prove that $G^* \cdot L = G'$.

Then we check that $L \cdot F' = F^*$. For a typical object A of B and morphism f: A → B in B, we have $L \cdot F'(\mathrm{A}) = [G'F'(\mathrm{A}), G'\nu'F'(\mathrm{A})] = [H(\mathrm{A}), \delta(\mathrm{A})] = F^*(\mathrm{A})$; while $L \cdot F'(f) = G'F'(f) = H(f) = F^*(f)$. Thus this relation is also proved. The proof of the theorem is complete. □

REMARKS. In fact, Pareigis, [27], proves also that the functors K, L defined in this proof are even unique. (See the exercises at the end of this chapter for a proof.)

7.8 WEAK ADJOINTS

The various types of semilattices and lattices can be treated as algebras or as ordered sets with closure for certain order-based operations. When these are considered as ordered algebras we get some inclusion functors that almost, but do not quite, have a left-adjoint. This leads us to the consideration of a generalization of the adjoint—the weak-adjoint of a functor, which is only a feeble functor. We look first at a typical problem of this sort.

In the category O of ordered sets we have a subcategory whose objects are complete lattices; in fact, several categories can be built with this class of objects. We note three: CL_1, with objects as complete lattices and morphisms as all monotone maps; CL_2, the same class of objects, but morphisms as maps that preserve arbitrary lattice sums; CL_3, with objects as before, but the morphisms as maps that preserve arbitrary lattice sums and lattice products. CL_1 is a full subcategory of O, CL_2 a subcategory of CL_1, CL_3 a subcategory of CL_2. We denote by J_i: $CL_i \to O$ ($i = 1, 2, 3$) the inclusion functors for these subcategories.

Let us consider as an object of O the (discrete) ordered set P = [(a, b, c), =]. We have objects C*, C#, and C′ from the CL_i, as in Figure 7.12. The monotone maps f^*: P → C*, $f\#$: P → C#, and f': P → C′ carry the elements a, b, c to (a^*, b^*, c^*), $(a\#, b\#, c\#)$, and (a', b', c'), respectively. We then have the following results regarding various types of paths at P:

(a) (f', C′) is a shortest J_3-path at P;
(b) ($f\#$, C#) is a shortest J_2-path at P;
(c) there is no shortest J_1-path at P.

PROOF. (a) If (g, C) is any J_3-path at P, we have a morphism g': C′ → C (in C_3) that takes a', b', c' to $g(a), g(b), g(c)$, respectively, and is uniquely defined for the other elements of C′ since these are all expressible as sums or products of the a', b', c' and g' is to preserve lattice sums and products. Clearly $g' \cdot f' = g$. This proves that (f', C′) is a shortest J_3-path at P.

(b) If (g, C) is any J_2-path at P, a morphism $g\#$: C# → C (in C_2) can be defined by setting the images of $a\#, b\#, c\#$ to be $g(a), g(b), g(c)$ and then extending $g\#$ as a lattice-sum-preserving map to all the other elements of C#. All these are expressible as sums from $a\#, b\#, c\#$ (even $O\#$ is the sum of the null subset of $(a\#, b\#, c\#)$). Then, $g\# \cdot f\# = g$, and ($f\#$, C#) is seen to be a shortest J_2-path at P.

(c) We shall see later that (f^*, C*) is what is called a canonical J_1-path at P. But it is not a shortest J_1-path, for (f'', C″) is a J_1- path at P such that there are two different morphisms g'', g''' from C* to C″, which differ in taking 1* to 1″ and 1‴, respectively, such that they satisfy $g'' \cdot f^* =$

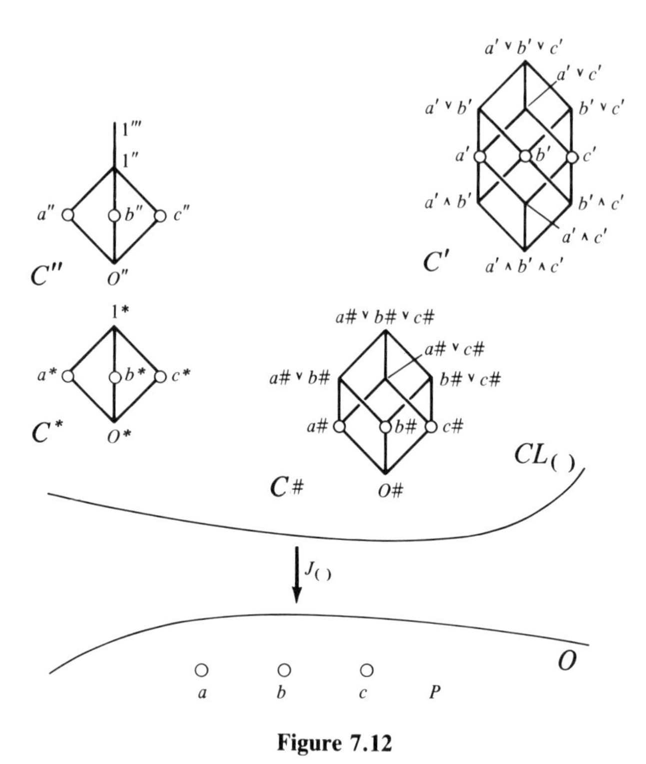

Figure 7.12

$f'' = g''' \cdot f^*$. A later result asserts that P cannot have a shortest J_1-path when it has a canonical J_1-path that is not a shortest J_1-path. This proves (c). □

So we now define the weaker form, called the *canonical G-path* for an object of B when G: $A \to B$ is a given functor; every shortest G-path would be a canonical G-path, not always the other way round.

Definition 7.8.1 Given the functor G: $A \to B$ and an object B of B, we say that:

(i) a G-path (f, A) at B is a *least G-path* at B if for every G-path (f', A') at B there is a path morphism g: $(f, \mathrm{A}) \to (f', \mathrm{A}')$, that is, a morphism g: $\mathrm{A} \to \mathrm{A}'$ in A such that $G(g) \cdot f = f'$;
(ii) a G-path (f, A) is a *canonical G-path* at B if it is a least G-path at B, and further [h: $\mathrm{A} \to \mathrm{A}$ is a morphism in A with $G(h) \cdot f = f$] implies that [$h = 1_{\mathrm{A}}$];
(iii) a G-path (f, A) is a *shortest G-path* at B if for every G-path (f', A') at B there is a *unique* path morphism g: $(f, \mathrm{A}) \to (f', \mathrm{A}')$ such that $G(g) \cdot f = f'$:

(iv) two G-paths (f, A) and (f', A') at B are *isomorphic* if there is a path morphism $g: (f, \text{A}) \to (f', \text{A}')$ that is an isomorphism in A.

Lemma 7.8.1 *(a) Given the functor $G: A \to B$ and a G-path (f, A) at an object* B *of B, [(f, A) is a shortest G-path] implies that [(f, A) is a canonical G-path], which implies that [(f, A) is a least G-path].*

(b) Given two G-paths (f, A) and (f', A') at B *that are isomorphic, if either of them is a shortest G-path or a canonical G-path or a least G-path, then so is the other.*

(c) If (f, A) and (f', A') are G-paths at B, *they are isomorphic if they are both canonical G-paths.*

PROOF. (a) When (f, A) is a shortest G-path, the uniqueness statement for the morphism $g: (f, \text{A}) \to (f', \text{A}')$ when applied to the G-path $(f', \text{A}') = (f, \text{A})$ implies that there is just one morphism $g: \text{A} \to \text{A}$ for which $G(g) \cdot f = f$. Since 1_A is surely one such g, it must be the only one. Thus the shortest G-path is a canonical G-path. The definition implies that a canonical G-path is a least G-path.

(b) This part follows quite easily from the definitions, since an isomorphism in A has an inverse.

(c) If (f, A) and (f', A') are both canonical G-paths at B, then from their definition it follows that there are path morphisms $g: (f, \text{A}) \to (f', \text{A}')$ and $g': (f', \text{A}') \to (f, \text{A})$ such that $G(g) \cdot f = f'$, $G(g') \cdot f' = f$. Hence $G(g' \cdot g) \cdot f = G(g') \cdot G(g) \cdot f = G(g') \cdot f' = f$, from which follows that $g' \cdot g = 1_\text{A}$; similarly, we get $G(g \cdot g') \cdot f' = f'$ and $g \cdot g' = 1_{\text{A}'}$. These together mean that g' is an inverse of g in A and so g is an isomorphism in A. Thus the two paths are isomorphic. □

REMARK. If there is a canonical G-path (f, A) and a shortest G-path (f', A'), both at B, then these are isomorphic G-paths, so both are shortest G-paths. So we get the following:

Corollary *If an object* B *of B has a canonical G-path (f,A) that is not a shortest G-path at* B, *then* B *cannot have a shortest G-path at all.*

Going back to the example given at the beginning of the section, we see quite easily that P has (f^*, C^*) as a canonical J_1-path; and we saw it was not a shortest J_1-path at P, so this P cannot have a shortest J_1-path at all.

Before analyzing the conditions under which each object of B would have a canonical G-path for a given functor $G: A \to B$, we treat two typical such examples using the constructions due to MacNeille ([23], Sections 11, 12).

Let O be the category of ordered sets and their monotone maps; by M we denote the subcategory of O whose objects are multiplicative semilattices and their product-preserving maps; thus $(X, \leq)$ is an object of M if every finite subset of M has a lattice product or greatest lower bound in M; and $f: (X, \leq) \to (Y, \leq)$ is a morphism in M when for any x_1, x_2 of X, $f(x_1 \wedge x_2) = f(x_1) \wedge f(x_2)$, $\wedge$ denoting the lattice multiplication.

For $(X, \leq)$ from $O(M)$, if $[x_i]$ is a subset of X, we set $x = \Sigma^*[x_i]$ and call x a "distributive sum" of the set $[x_i]$ if x is a lattice sum (or least upper bound) of $[x_i]$ and further $(b \wedge x)$ is a lattice sum of $[b \wedge x_i]$ for each element b of X. A map $f: X \to Y$ is said to preserve distributive sums from $(X, \leq)$ to $(Y, \leq)$ (objects of M) if $[x = \Sigma^*[x_i]$ in $(X, \leq)]$ implies that $[f(x) = \Sigma^* [f(x_i)]$ in $(Y, \leq)]$. We take M^* to be that subcategory of M that has the same objects as M but whose morphisms are maps preserving finite products and distributive sums. The full subcategory of M^* whose objects are complete lattices in which all sums are distributive is denoted by DS. We then have the following results:

Theorem 7.8.1 *(a) If $J: C \to O$ is the inclusion functor, each object $(X, \leq)$ of O has a canonical J-path $(p, (X^*, \subseteq))$; the complete lattice $(X^*, \subseteq)$ is called the canonical completion of the ordered set $(X, \leq)$.*

(b) If $K: DS \to M^*$ is the inclusion functor, each object $(X, \leq)$ of M^* has a shortest K-path $(p, (X\#, \subseteq))$.

PROOF. (a) Given the ordered set $(X, \leq)$, we first associate to each element x of X the initial set $p(x) = [y$ in X: $y \leq x]$, which is called the *principal ideal* of $(X, \leq)$ defined by x. Then for any subset A of X we set $\text{comp}(A) = [\Pi p(x)$: $p(x)$ contains $A]$; if there is no $p(x)$ that contains A, the intersection of this null family would give $\text{comp}(A) = X$. Calling a subset B of X a *comprincipal ideal* of $(X, \leq)$ if $B = [\Pi p(x)$: $p(x)$ contains $B]$, we see that $\text{comp}(A)$ is a comprincipal ideal for any subset A of X, and that it is the smallest comprincipal ideal containing A; and for a one-element set $[x]$, $\text{comp}[x] = p(x)$. We take X^* to be the family of all comprincipal ideals of $(X, \leq)$. Under the order $\subseteq$, it is clear that $(X^*, \subseteq)$ is a complete lattice; for a subfamily $[B_j]$ from X^*, a lattice product of the family is the intersection $\Pi(B_j)$, while a lattice sum for this subfamily is comp[union of the B_j]. We have an injective monotone map p of $(X, \leq)$ in $(X^*, \subseteq)$, where $p(x)$ is the principal ideal defined by x. Evidently $(p, (X,^* \subseteq))$ is a J-path at $(X, \leq)$. For any J-path $(f, (Y, \leq))$ at $(X, \leq)$, we have an f^*: $(X, \subseteq) \to (Y, \leq)$ given by $f^*(B) =$ the lattice product $\wedge[f(x)$: for all x of X for which $p(x)$ contains $B]$. Surely this is a monotone map, and $f^*(p(x)) = f(x)$ for any x of X. That is, $(p, (X^*, \subseteq))$ is a least J-path at (X, F). To check then that it is a canonical J-path, let h: $(X^*, \subseteq) \to (X^*, \subseteq)$ be a monotone map such that $h(p(x)) = p(x)$ for each x of X. Then for any B of X^*, $B = \Pi[p(x)$: for all $p(x)$ containing $B]$ implies that $h(B) \subseteq h(p(x))$ for all $p(x)$ containing B, or $h(B) \subseteq$ each $p(x)$ containing B, hence $h(B) \subseteq$ their intersection $= B$; so we have $h(B) \subseteq B$. On the other hand, if $y \in B$, $p(y) \subseteq B$, so $h(p(y)) \subseteq h(B)$; that is, $p(y) \subseteq h(B)$. But y belongs to $p(y)$, so that y is in $h(B)$ too, or $B \subseteq h(B)$. These prove that $h(B) = B$ for each B of X^*, or $h = 1_{X^*}$. The path $(p, (X^*, \subseteq))$ is a canonical J-path at $(X, \leq)$.

(b) If $(X, \leq)$ is a multiplicative semilattice (an object of M^*), we set $X\#$ = [the family of Σ^*-ideals of $(X, \leq)$], where a subset B of X is called a Σ^*-ideal if it is an initial set in $(X, \leq)$ that contains all existing distributive sums of sets of elements chosen from B as well as finite products of elements chosen from B. For an x of X it is true that $p(x)$ is a Σ^*-ideal. Any family (B_j) of Σ^*-ideals has a lattice sum in $(X\#, \subseteq)$, namely, the least Σ^*-ideal containing the union of the B_j [equal to those elements of X that are expressible as distributive sums in $(X, \leq)$ of subsets of the union of the B_j]; and the family (B_j) has a lattice product, namely, the intersection of the B_j. Further, the sums in $(X\#, \subseteq)$ are all distributive. This can be verified from the way the sum was formed for a family of the ideals. Hence $(p, (X\#, \subseteq))$ is a K-path at $(X, \leq)$. If $(g, (Y, \leq))$ is any K-path at $(X, \leq)$, we note first that an element B of $X\#$ can be expressed as $\Sigma^*[p(x)$: x in $B]$ in $(X\#, \subseteq)$. We then define the map $g\#: X\# \to Y$ by setting $g\#(B) = \Sigma^*[g(x)$: for all x in $B]$. It is not hard to show that this $g\#$ is a map that preserves finite lattice products and distributive sums from $(X, \leq)$ to $(Y, \leq)$; and $g\# \cdot p = g$. That any morphism g': $(X\#, \subseteq) \to (Y, \leq)$ in DS for which $g' \cdot p = g$ must be the same as $g\#$ is clear from the remark about the representation of an element of $X\#$ as a distributive sum. Thus $(p, (X\#, \subseteq))$ is a shortest K-path at $(X, \leq)$. □

We consider next compositions and resolutions of these new path types.

Lemma 7.8.2 *Let G: $A \to B$ and H: $B \to C$ be given functors, and* C *an object of C.*

(i) If $(f$, B*) is a least H-path at* C *and $(g$,* A*) is a least G-path at* B, *then $(H(g) \cdot f$,* A*) is a least $(H \cdot G)$-path at* C;

(ii) if $(f$, B*) is a shortest H-path at* C *and $(g$,* A*) is a canonical G-path at* B, *then $(H(g) \cdot f$,* A*) is a canonical $(H \cdot G)$-path at* C;

(iii) if $(f$, B*) is a shortest H-path at* C *and $(h$,* A*) is a least or a canonical $(H \cdot G)$-path at* C, *then there is a least or a canonical G-path $(g$,* A*) at* B *such that $h = H(g) \cdot f$.*

PROOF. (i) From the hypotheses it follows that $(H(g) \cdot f$, A) is an $(H \cdot G)$-path at C. If $(k$, A$')$ is any $(H \cdot G)$-path at C, then $(k, G($A$'))$ is an H-path at C; hence there is an m: B $\to G($A$')$ such that $H(m) \cdot f = k$. Then $(m$, A$')$ is a G-path at B, so that there is an n: A $\to$ A$'$ such that $G(n) \cdot g = m$.

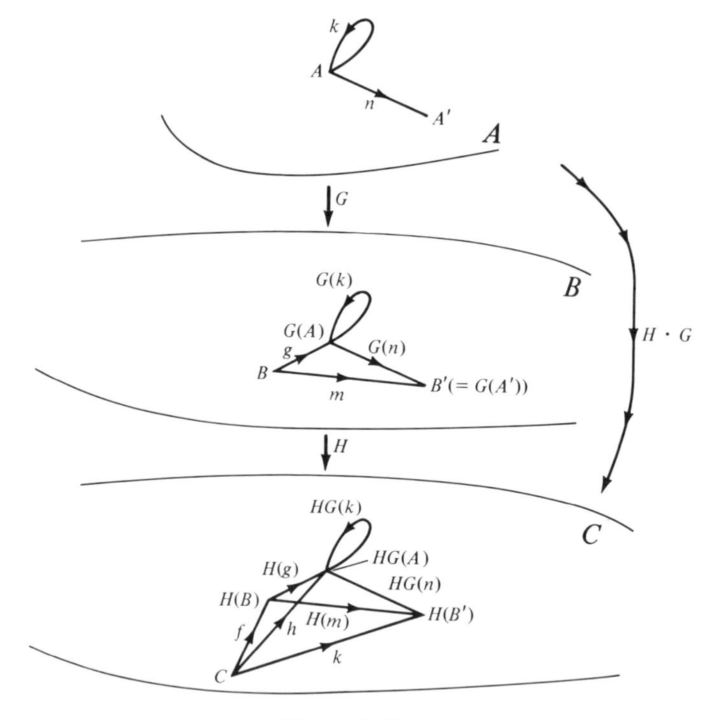

Figure 7.13

Then we have $(H \cdot G) \cdot (n) \cdot H(g) \cdot f = H(G(n) \cdot g) \cdot f = H(m) \cdot f = k$, proving that $(G(n) \cdot g$, A) is a least $(H \cdot G)$-path at C. (See Figure 7.13.)

(ii) From the last result it already follows from the hypotheses that $(H(g) \cdot f$, A) is a least $(H \cdot G)$-path at C. To check that it is canonical, we proceed as follows. If for some k: A → A we had $(H \cdot G) \cdot (k) \cdot H(g) \cdot f = H(g) \cdot f$, from $H(G(k) \cdot g) \cdot f = H(g) \cdot f$ and the fact that (f, B) is a shortest H-path it would follow that $G(k) \cdot g = g$. Then, since (g, A) is a canonical G-path, $k = 1_A$ follows. This completes the proof.

(iii) We take first the case when (h, A) is a least $(H \cdot G)$-path at C, while (f, B) is a shortest H-path at C. Since (h, G(A)) is an H-path at C, there is a unique g: B → G(A) such that $h = H(g) \cdot f$. This makes (g, A) a G-path at B. To show that it is a least such path, let (m, A′) be any G-path at B. Since $(H(m) \cdot f$, A′) is then an $(H \cdot G)$-path at C, there is an n: A → A′ such that $(H \cdot G)(n) \cdot h = H(m) \cdot f$, or $H(G(n) \cdot g) \cdot f = H(m) \cdot f$. Since ($f$, B) is a shortest G-path at C, it would follow that $G(n) \cdot g = m$, which really proves that (g, A) is a least G-path at B. If we next assume that (h, A) is even a canonical $(H \cdot G)$-path at C, as before we see that (g, A) is a least G-path at B. To see that it is canonical, we note that if k: A $\to$ A satisfies $G(k) \cdot g = g$, then $(H \cdot G) \cdot (k) \cdot H(g) \cdot f = H(g) \cdot f$ or $(H \cdot G)(k) \cdot h = h$, from which we get $k = 1_{\mathrm{A}}$. □

To get analogues of Theorems 7.2.1 and 7.3.1 for the least or canonical paths, we start by defining the weak adjoint.

Definition 7.8.2 If G: $A \to B$ is a functor and F: $B \dashrightarrow A$ is a feeblc functor, and CR denotes the category of sets and cofull relations between sets, there are two functors $\hom_A(F \times 1_A)$ and $\hom_B(1_B \times G)$, both from $B^{\mathrm{op}} \times A$ to CR: the first is composition of $F \times 1_A$: $B^{\mathrm{op}} \times A \to A^{\mathrm{op}} \times A$ with $\hom_A$: $A^{\mathrm{op}} \times A \to CR$, while the second is the composite of $1_B \times G$; $B^{\mathrm{op}} \times A \to B^{\mathrm{op}} \times B$, followed by $\hom_B$: $B^{\mathrm{op}} \times B \to CR$.

Definition 7.8.3 For a functor G: $A \to B$, the feeble functor F: $B \dashrightarrow A$ is called a *weak left-adjoint* of G if there is a natural transformation n: $\hom_A(F \times 1_A) \to \hom_B(1_B \times G)$ and a left-natural transformation m: $\hom_B(1_B \times G) \to \hom_A(F \times 1_A)$ such that $n \circ m = [1_{\hom_B(1_B \times G)}]$, while, $m \circ n$ contains $1_{\hom_A(F \times 1_A)}$. For simplicity in the following we shorten $\hom_A(F \times 1_A)$ to $h(F \times 1)$ and similarly $\hom_B(1_B \times G)$ to $h(1 \times G)$.

Theorem 7.8.2 *Given a functor G: $A \to B$, conditions (i), (ii), and (iii) following are equivalent, and conditions (i*), (ii*), (iii*) are equivalent:*

- *(i) there is a feeble functor F: $B \dashrightarrow A$ that is a weak left-adjoint of G;*
- *(i*) besides (i), if (n, m) are the natural transformations defining the weak left-adjointness, [B $\in O(B)$, $h \in \hom_A(F(\mathrm{B}), F(\mathrm{B}))$, $(1_{F(\mathrm{B})}, k) \in n(\mathrm{B}, F(\mathrm{B}))$ and $G(h) \cdot k = k$] imply that $[h = 1_{F(\mathrm{B})}]$;*
- *(ii) there is a feeble functor F: $B \dashrightarrow A$ and a natural transformation η: $1_B \to G \cdot F$ and, for each A of $O(A)$, a nonnull set $\nu(\mathrm{A})$ of morphisms from $\hom_A(FG(\mathrm{A}), \mathrm{A})$ such that (a) for all A of $O(A)$, $G(\nu(\mathrm{A})) \cdot \eta(G(\mathrm{A})) = 1_{G(\mathrm{A})}$, and (b) for objects A, B from A, B, $[f$ in $\hom_B(\mathrm{B}, G(\mathrm{A}))$, $h \in \nu(\mathrm{A}) \cdot F(f)] \Rightarrow [G(h) \cdot \eta(\mathrm{B}) = f]$;*
- *(ii*) add to the conditions in (ii): (c) for B of $O(B)$, $[h \in \hom_A(F(\mathrm{B}), F(\mathrm{B}))$, $G(h) \cdot \eta(\mathrm{B}) = \eta(\mathrm{B})] \Rightarrow [h = 1_{F(\mathrm{B})}]$;*
- *(iii) each object B of B has a least G-path;*
- *(iii*) each object of B has a canonical G-path.*

PROOF. We show that (i) $\Rightarrow$ (ii) $\Rightarrow$ (iii) $\Rightarrow$ (i), and (i*) $\Rightarrow$ (ii*) $\Rightarrow$ (iii*) $\Rightarrow$ (i*).

(i) $\Rightarrow$ (ii). Assuming that we have a natural transformation n: $h(F \times 1) \to h(1 \times G)$ and a left-natural transformation m: $h(1 \times G) \to h(F \times 1)$, with $n \circ m = (1_{h(1 \times G)})$ and $m \circ n$ containing $1_{h(F \times 1)}$ where F: $B \dashrightarrow A$ is a feeble functor, the conditions on $n \circ m$ and $m \circ n$ imply the following: for

each f in $\hom_A(F(\mathrm{B}), \mathrm{A}))$ (for any objects A, B from A, B) there is a g in $\hom_B(\mathrm{B}, G(\mathrm{A}))$ such that

$$(f, g) \in n(\mathrm{B}, \mathrm{A}) \quad \text{and} \quad (g, f) \in m(\mathrm{B}, \mathrm{A}); \tag{7.8.1}$$

for each g in $\hom_B(\mathrm{B}, G(\mathrm{A}))$ there is an f in $\hom_A(F(\mathrm{B}), \mathrm{A})$ such that

$$(g, f) \in m(\mathrm{B}, \mathrm{A}) \quad \text{and} \quad (f, g) \in n(\mathrm{B}, \mathrm{A}); \tag{7.8.2}$$

for an f of $\hom_A(F(\mathrm{B}), \mathrm{A})$ and for a g in $\hom_B(\mathrm{B}, G(\mathrm{A}))$ as in (7.8.1), if there is a g' in $\hom_B(\mathrm{B}, G(\mathrm{A}))$ such that

$$(f, g') \in n(\mathrm{B}, \mathrm{A}), \quad \text{then} \quad g' = g. \tag{7.8.3}$$

The net result of these three statements is that $n(\mathrm{B}, \mathrm{A})$ is effectively a surjective mapping of $\hom_A(F(\mathrm{B}), \mathrm{A})$ on $\hom_B(\mathrm{B}, G(\mathrm{A}))$. Taking $\mathrm{A} = F(\mathrm{B})$ in (7.8.1) and $f = 1_{F(\mathrm{B})}$, we then see that there is a unique morphism $\eta(\mathrm{B})$ in $\hom_B(\mathrm{B}, GF(\mathrm{B}))$ such that $(1_{F(\mathrm{B})}, \eta(\mathrm{B}))$ is in $n(\mathrm{B}, F(\mathrm{B}))$ and $(\eta(\mathrm{B}), 1_{F(\mathrm{B})})$ is in $m(\mathrm{B}, F(\mathrm{B}))$; and taking $\mathrm{B} = G(\mathrm{A})$ and $g = 1_{G(\mathrm{A})}$ in (7.8.2), we get a set of morphisms h of $\hom_A(FG(\mathrm{A}), \mathrm{A})$ for which $(1_{G(\mathrm{A})}, h) \in m(G(\mathrm{A}), \mathrm{A})$ and $(h, 1_{G(\mathrm{A})}) \in n(G(\mathrm{A}), \mathrm{A})$. We take this nonnull set to be $\nu(\mathrm{A})$. We have to check conditions (a), (b).

First we check that η is a natural transformation from 1_B to GF: given a typical morphism b: $\mathrm{B}' \to \mathrm{B}$ in B, we consider in $B^{\mathrm{op}} \times A$ the morphisms $(b, 1)$: $(\mathrm{B}, F(\mathrm{B})) \to (\mathrm{B}', F(\mathrm{B}))$ and $(1, k)$: $(\mathrm{B}', F(\mathrm{B}')) \to (\mathrm{B}', F(\mathrm{B}))$, where k is any morphism from $F(b)$. Using the naturalness of n from $h(F \times 1)$ to $h(1 \times G)$, we get (see Figure 7.14) that $n(\mathrm{B}', F(\mathrm{B}))$ contains $[1_{F(\mathrm{B})} \cdot 1_{F(\mathrm{B})} \cdot F(b), G(1_{F(\mathrm{B})}) \cdot \eta(\mathrm{B}) \cdot b]$; and it also contains (from the right square) $[k \cdot 1_{F(\mathrm{B}')} \cdot F(1_{\mathrm{B}'}), G(k) \cdot \eta(\mathrm{B}') \cdot 1_{\mathrm{B}'}]$. Thus for any k of $F(b)$ we have both $(k, \eta(\mathrm{B}) \cdot b)$ and $(k, G(k) \cdot \eta(\mathrm{B}'))$ in $n(\mathrm{B}', F(\mathrm{B}))$; since we saw that $n(\mathrm{B}, \mathrm{A})$ is always a mapping relation, it follows that $\eta(\mathrm{B}) \cdot b = G(k) \cdot \eta(\mathrm{B}')$. This

Figure 7.14

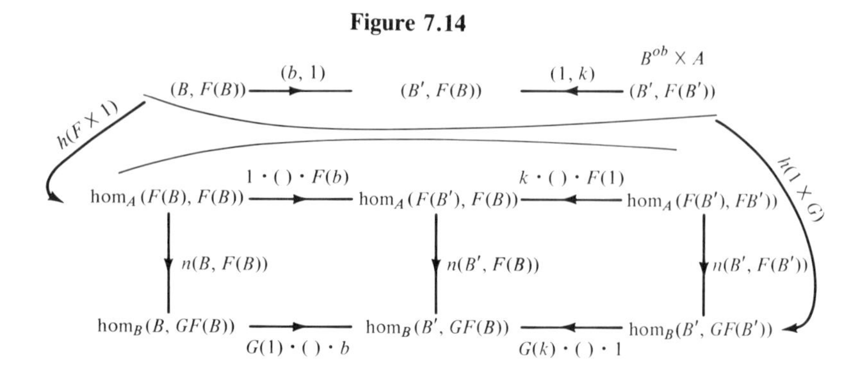

CR

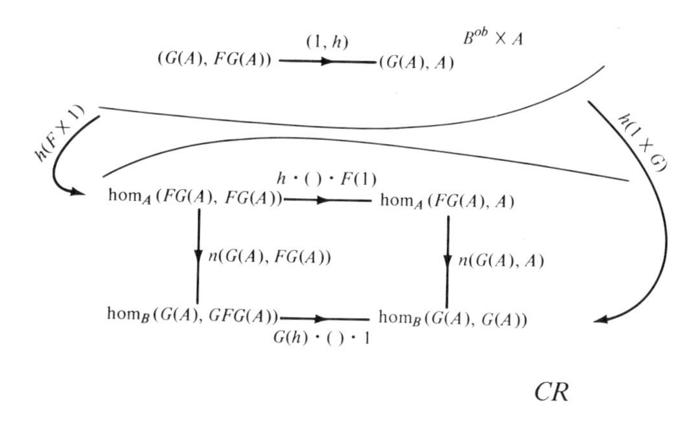

Figure 7.15

means that $G(F(b)) \cdot \eta(B') = \eta(B) \cdot b$, which makes η a natural transformation.

To prove (a), let A be an object of A and h any morphism from $\nu(A)$ (see Figure 7.15). In $B^{op} \times A$ we consider the morphism $(1, h)$: $(G(A), FG(A)) \to (G(A), A)$; using the naturalness of n and following up the two paths from the element $1_{FG(A)}$ of $\hom_A(FG(A), FG(A))$, we get $n(G(A), A)$ contains $[h \cdot 1_{FG(A)} \cdot F(1_{G(A)}), G(h) \cdot \eta(G(A)) \cdot 1_{G(A)}]$. Since $n(G(A), A)$ also contains $(h, 1_{G(A)})$, by the choice of h, (and since $n(G(A), A)$ is a mapping relation), we deduce that $G(h) \cdot \eta(G(A)) = 1_{G(A)}$ for each h of $\nu(A)$; thus $G(\nu(A)) \cdot \eta(G(A)) = 1_{G(A)}$.

Next, to prove (b) let f be a morphism from $\hom_B(B, G(A))$ and let h be from $\nu(A) \cdot F(f)$ and therefore of the form $h = h_1 \cdot k_1$ with $h_1 \in \nu(A)$, $k_1 \in F(f)$; then $G(h) \cdot \eta(B) = G(h_1) \cdot G(k_1) \cdot \eta(B) = G(h_1)\eta(G(A)) \cdot f$ (since η is natural) $= 1_{G(A)} \cdot f$ [because of the last result in (a)] $= f$. This proves (b).

(i) ⇒ (ii*).* Condition (c) of (ii*) follows from the extra condition in (i*), since we defined $\eta(B)$ as the unique morphism in $\hom_B(B, GF(B))$ for which $(1_{F(B)}, \eta(B))$ is in $n(B, F(B))$.

(ii) ⇒ (iii). Assuming (ii), it is clear that for any B of $O(B)$, $(\eta(B), F(B))$ is a G-path at B; we check now that it is a least such G-path. For if (f, A) is any G-path at B, so that f: B $\to G(A)$, we know that $\nu(A)$ and $F(f)$ are nonnull and so is $\nu(A) \cdot F(f)$. If h is any morphism from this $\nu(A) \cdot F(f)$, then h is in $\hom_A(F(B), A)$ and by part (b) in (ii), $G(h) \cdot \eta(B) = f$. This proves that $(\eta(B), F(B))$ is a least G-path at B.

(ii) ⇒ (iii*).* That $(\eta(B), F(B))$ is also a canonical G-path if (c) of (ii*) is assumed follows easily enough.

(iii) $\Rightarrow$ *(i)*. Assume that there is a least G-path, which we choose as $(\eta(\mathrm{B}), F(\mathrm{B}))$ for each B of $O(B)$. We extend the F so defined for objects of B to morphisms by setting $F(b)$, for a b: $\mathrm{B}' \to \mathrm{B}$ in $M(B)$ to be the set of morphisms h: $F(\mathrm{B}') \to F(\mathrm{B})$ for which $G(h) \cdot \eta(\mathrm{B}') = \eta(\mathrm{B}) \cdot b$; such morphisms h exist since $(\eta(\mathrm{B}'), F(\mathrm{B}'))$ is a least G-path at B', while $(\eta(\mathrm{B}) \cdot b, F(\mathrm{B}))$ is a G-path at B' also. That this F gives a feeble functor from B to A follows when we note that $1_{F(\mathrm{B})}$ is surely in $F(1_{\mathrm{B}}$: $\mathrm{B} \to \mathrm{B})$, and if b': $\mathrm{B}'' \to \mathrm{B}'$, b: $\mathrm{B}' \to \mathrm{B}$ are both morphisms of B, and h', h are from $F(b')$, $F(b)$, respectively, then $G(h \cdot h') \cdot \eta(\mathrm{B}'') = G(h) \cdot G(h') \cdot \eta(\mathrm{B}'') = G(h) \cdot \eta(\mathrm{B}') \cdot b' = \eta(\mathrm{B}) \cdot b \cdot b'$, so that $h \cdot h'$ is in $F(b \cdot b')$, or $F(b) \cdot F(b')$ is contained in $F(b \cdot b')$. Thus F is a feeble functor.

We then define the natural transformations n and m; first recall that for a typical morphism (b, a): $(\mathrm{B}, \mathrm{A}) \to (\mathrm{B}', \mathrm{A}')$ of $B^{\mathrm{op}} \times A$, $h(F \times 1)(b, a)$ is a cofull relation from $\hom_A(F(\mathrm{B}), \mathrm{A})$ to $\hom_B(\mathrm{B}, G(\mathrm{A}))$ such that (f, f') is in $h(F \times 1)(b, a)$ iff f' is in $a \cdot f \cdot F(b)$; and $h(1 \times G)(b, a)$ is a cofull relation from $\hom_B(\mathrm{B}, G(\mathrm{A}))$ to $\hom_A(F(\mathrm{B}), \mathrm{A})$ consisting of those (g, g') for which g' is in $G(a) \cdot g \cdot b$. We now set $n(\mathrm{B}, \mathrm{A}) = [(f, g)$: f in $\hom_A(F(\mathrm{B}), \mathrm{A})$, g in $\hom_B(\mathrm{B}, G(\mathrm{A}))$, $g = G(f) \cdot \eta(\mathrm{B})]$ and $m(\mathrm{B}, \mathrm{A}) = [n(\mathrm{B}, \mathrm{A})]^{\mathrm{r}}$, the reverse of the relation $n(\mathrm{B}, \mathrm{A})$. Obviously $n(\mathrm{B}, \mathrm{A})$ is cofull; $m(\mathrm{B}, \mathrm{A})$ is also cofull, for if g is from $\hom_B(\mathrm{B}, G(\mathrm{A}))$, (g, A) would be a G-path at B. Since $(\eta(\mathrm{B}), G(\mathrm{B}))$ is a least G-path at B, there is an f in $\hom_A(F(\mathrm{B}), \mathrm{A})$ such that $G(f) \cdot \eta(\mathrm{B}) = g$, which is the same as saying that (f, g) is in $n(\mathrm{B}, \mathrm{A})$ and (g, f) is in $m(\mathrm{B}, \mathrm{A})$. The verifications that $n \circ m = [1_{h(1 \times G)}]$ and $m \circ n$ contains $1_{h(F \times 1)}$ are not difficult.

There is then the need to check that n is a natural transformation and m is a left-natural transformation, to complete the proof that F is a weak left-adjoint of G. For n to be natural we should show that for a typical (b, a) of $M(B^{\mathrm{op}} \times A)$, $[h(1 \times G)(b, a)] \circ [n(\mathrm{B}, \mathrm{A})] = [n(\mathrm{B}', \mathrm{A}')] \circ [h(F \times 1)(b, a)]$. (See Figure 7.16.) The pair (f, g') is from the left-hand side of the preceding equation iff [for some g in $\hom_B(\mathrm{B}, G(\mathrm{A}))$, $g = G(f) \cdot \eta(\mathrm{B})$ and $g' = G(a) \cdot g \cdot b$], that is, iff [$g'$ is in $G(a) \cdot G(f) \cdot \eta(\mathrm{B}) \cdot b$]; and a pair (f, g'') is from the right-hand side of the equation iff [for an f' of $\hom_A(F(\mathrm{B}'), \mathrm{A}')$, f' is in $a \cdot f \cdot F(b)$ and $g'' = G(f') \cdot \eta(\mathrm{B}')$], that is, iff [$g''$ is in $G(a \cdot f \cdot F(b)) \cdot \eta(\mathrm{B}')$]. Since the naturalness of η implies that $GF(b) \cdot \eta(\mathrm{B}') = \eta(\mathrm{B}) \cdot b$, it follows that the two sides of the equation consist of the same pairs. So n is natural. To prove that m is left natural we have to show, for a typical (b, a), that $[h(F \times 1)(b, a)] \circ [m(\mathrm{B}, \mathrm{A})]$ is contained in $[m(\mathrm{B}', \mathrm{A}')] \circ [h(1 \times G)(b, a)]$. A pair (g, f') is from $[h(F \times 1)(b, a)] \circ [m(\mathrm{B}, \mathrm{A})]$ iff [for an f of $\hom_A(F(\mathrm{B}), \mathrm{A})$, (g, f) is in $m(\mathrm{B}, \mathrm{A})$ and (f, f') is in $h(F \times 1)(b, a)$]; that is, iff [for an f, $G(f) \cdot \eta(\mathrm{B}) = g$ and f' is in $a \cdot f \cdot F(b)$]. This pair would be in $[m(\mathrm{B}', \mathrm{A}')] \circ [h(1 \times g)(b, a)]$ provided that for a g' of $\hom_B(\mathrm{B}', G(\mathrm{A}'))$, (g, g') is in $h(1 \times G)(b, a)$ and (g', f') is in $m(\mathrm{B}', \mathrm{A}')$; but if we set $g' = G(f') \cdot \eta(\mathrm{B}')$, this g' does satisfy the two conditions. Thus m is a left-natural transformation. These prove that (iii) implies (i).

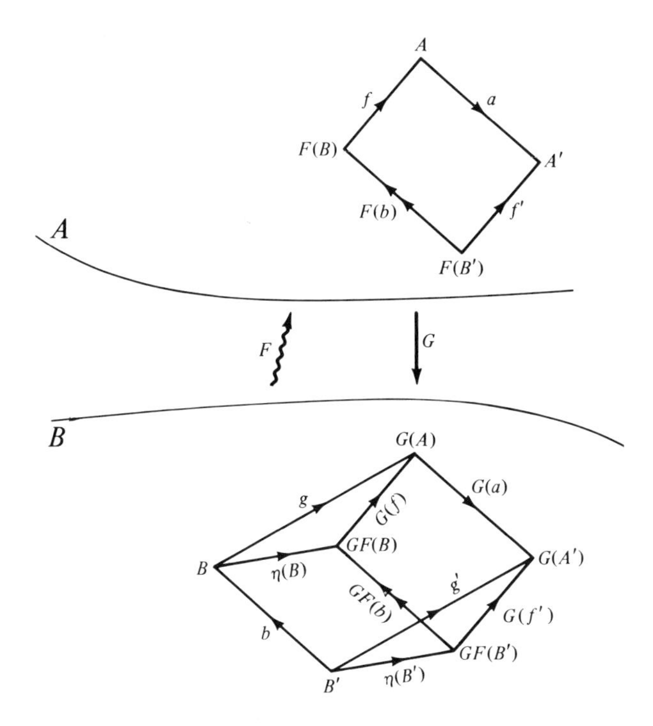

Figure 7.16

(iii) ⇒ (i*).* The extra assumption in the preceding proof that (η(B), G(B)) is a canonical G-path at B would mean that [$G(h) \cdot \eta(\text{B}) = \eta(\text{B})$ for an h in $\hom_A(F(\text{B}), F(\text{B}))$] implies that [$h = 1_{F(\text{B})}$], which added to (i) gives (i*).

Since (i*) is equivalent to (iii*), we shall say that a feeble functor F: B--->A is a *canonical weak left-adjoint* of G: $A \to B$ if (i*) is true. Thus the equivalence of (i*) and (iii*) now reads: G: $A \to B$ has a canonical weak left-adjoint iff each object of B has a canonical G-path. We then get a corollary to Theorem 7.8.1, part (a): □

Corollary *The inclusion functor J: C → O has a canonical weak left-adjoint C*: O → C; this C* associates to each ordered set its canonical completion.*

EXERCISES AND REMARKS ON CHAPTER 7

1. Left-adjoints among functors between *FD*-algebras.

   We consider the categories *OS, Sg, OSg, Sgu, OSgu, Gp,* and *ISg*, whose objects are, in order, orthocomplemented sets, semigroups, orthocomplemented semigroups, semigroups with unit, orthocomplemented semigroups with unit, groups, and inverse semigroups. All these can be described as FD-algebras for a suitable choice of F from the three operations $[(\cdot, 2), ((\ )', 1), (u, 0)]$; the first is binary, the second unary, and the third nullary. The D is a suitable selection of pairs from the following set of pairs of elements from the free F-algebra over a three-element set $A = [a, b, c]$:

$$p_1 = [(a \cdot b) \cdot c; a \cdot (b \cdot c)]; p_2 = [((a'))'; a]; p_3 = [(a \cdot b)'; (b)' \cdot (a)'];$$

$$p_4 = [a \cdot (a)' \cdot a; a]; p_5 = [(a \cdot (a)') \cdot (b \cdot (b)'); (b \cdot (b)') \cdot (a \cdot (a)')];$$

$$p_6 = [a \cdot (a)'; u]; p_7 = [a \cdot u; a]; p_8 = [u \cdot a; a].$$

Then an orthocomplemented set is an FD-algebra with $F = [((\ )', 1)]$, $D = [p_2]$; a semigroup is an FD-algebra with $F = [(\cdot, 2)]$, $D = [p_1]$; an orthocomplemented semigroup is an FD-algebra with $F = [(\cdot, 2), ((\ )', 1)]$, $D = [p_1, p_2, p_3]$; a semigroup or orthocomplemented semigroup with unit has $(u,0)$ added to the F and (p_7, p_8) added to the D; a group is an FD-algebra with F containing all three operations and $D = [p_1, p_2, p_3, p_6, p_7, p_8]$; while an inverse semigroup is an FD-algebra with $F = [(\cdot, 2), ((\ )', 1)]$ and $D = [p_1, p_2, p_3, p_4, p_5]$.

Among these categories we have the following inclusion or forgetful functors: J^*: $GP \to OSgu$, K^*: $OSgu \to Sgu$, B^*: $Sgu \to S$, J: $ISg \to OSg$, K: $OSg \to Sg$, B: $Sg \to S$, and V: $OS \to S$, where S is the category of sets and set maps. All of them have left-adjoints. Some of the associated shortest paths are described in what follows; the reader is to verify that they are indeed such shortest paths.

(a) For an object X in S, a shortest V-path is $[h, ((X \cup X'), (\ )^*)]$, where X' is a set disjoint with and equipotent with X, with an assigned bijection $x \leftrightarrow x'$ between X and X'. We set $(y)^* = x'$ or x, according as $y = x$ or x'; and h is the inclusion map of X in XUX'.

(b) For an X from $O(S)$, a shortest B-path at X is $[s, (W(X), \circ)]$, where the elements of $W(X)$ are "words" over X, that is, finite (nonnull) sequences of the form $x_1, \ldots, x_n$ with the x_i from X; and the composition of two such words is given by juxtaposition, that is, $(x_1, \ldots, x_n) \circ (y_1, \ldots, y_m) = x_1, \ldots, x_n, y_1, \ldots, y_m$. And s is the map of X in $W(X)$ that takes x of X to the single-letter word x.

(b*) For an X of $O(S)$, a shortest B^*-path at X is $[s, (W^*(X), \circ)]$; this time $W^*(X)$ contains, besides the words over X as in (b), the null word $\varnothing$; and the operator $\circ$ is such that $w \circ \varnothing = w = \varnothing \circ w$ for each w of $W^*(X)$. The mapping s is as in (b).

(c) For an $(X, \cdot)$ of $O(Sg)$, a shortest K-path is $[p \cdot s, (W(X \cup X'), \circ)/E]$; the word semigroup $((W(X \cup X'), \circ)$ is formed from $X \cup X'$ where X' is a set disjoint with X with the the same cardinality as X and with an assigned bijection $x \leftrightarrow x'$ between the two sets X and X'. Then we form the quotient semigroup of this by a congruence E defned by (w, w') is in E iff there is a finite sequence of words $w = w_0, w_1, \ldots, w_m = w'$ from w to w' in $W(X \cup X')$ such that the passage from a w_i to the next w_{i+1} in the sequence is by one of the following elementary operations: (i) replace a pair of consecutive letters in w_i of the form x, y by a single letter z when, in $(X, \cdot)$, $x \cdot y = z$; (i′) replace a single letter z in w_i by a pair of consecutive letters x, y when $x \cdot y = z$ in $(X, \cdot)$; (ii) replace a pair of consecutive letters in w_i of the form y', x' by a single letter z' when $z = x \cdot y$ in $(X, \cdot)$; (ii′) replace the single letter z' in w_i by the pair of consecutive ones y', x' when $z = x \cdot y$ in $(X, \cdot)$.

In this quotient semigroup the orthocomplementation $(\ )^*$ is defined by setting $[E(z_1, \dots, z_n)]^* = E(z_n^*, \dots, z_1^*)$ where $z^* = x'$ or x according as z (of $X \cup X'$) is an x or an x'. (Note that $E(x) = (p \cdot s)(x)$ for each x of X.)

(c*) For an $(X, \cdot, 1)$ from $O(Sgu)$, the same type of construction as in (c) leads us to a shortest K^*-path at $(X, \cdot, 1)$.

(d) For an object $(X, \cdot, (\ )')$ of OSg we have a shortest J-path $[p, (X, \cdot, (\ )')/E]$ where E is a congruence on $(X, \cdot, (\ )')$ and p is the canonical projection of X on X/E; we define E by $(x, y) \in E$ iff there is a finite sequence $x = x_0, \dots, x_m = y$ of elements of X such that for each successive pair (x_i, x_{i+1}) one of the following is true: (i) from a finite factorization of x_i [in $(X, \cdot)$] we get one for x_{i+1} by replacing a consecutive trio of factors of the form (x, x', x) by a single factor x [or a trio of the form (x', x, x') by a single x']; (i′) the reverse of relation (i), that is, x_i is obtainable from x_{i+1} as in (i); (ii) calling a pair of elements p, q of X an ortho pair if they are of the form x, x' or of the form x', x, there is a finite factorization of x_i in which there occur four consecutive terms of the form p, q, r, s with (p, q) and (r, s) both ortho pairs, and a corresponding factorization of x_{i+1} is obtained by rearranging the order of these four factors as r, s, p, q.

(d*) Again starting with an object $(X, \cdot. (\ )', 1)$ from $OSGu$, we have a shortest J^*-path at this object of the form $[p, (X, \cdot, (\ )', 1)/E]$ where the congruence E is now defined by $(x, y) \in E$ iff there is a finite sequence $x = x_0, \dots, x_m = y$ of elements of X from x to y such that successive elements x_i, x_{i+1} of the sequence satisfy one of the following conditions: (i) from a factorization of x_i in $(X, \cdot)$ we can get a factorization for x_{i+1} by replacing an ortho pair in the first factorization by a single factor 1; (i′) the reverse of (i); (ii) from a factorization of x_i one for x_{i+1} is obtained by adding a single extra factor 1 somewhere; (ii′) the reverse of (ii).

2. The *Malčev conditions* for a semigroup to be embeddable in a group.

Starting from any semigroup $(X, \cdot)$ we can find a shortest U-path at $(X, \cdot)$ for the inclusion functor U: $SGu \to Sg$. If $(X, \cdot)$ has a unit element, this is merely $(1_X, (X, \cdot))$ if $(X, \cdot)$ has no unit element, we consider $X \mathring{\cup} (1)$, the union of X with a new one-element set (1); and define in this set a binary multiplication that extends the one in X by setting $x \cdot 1 = x = 1 \cdot x$ for any x of X and $1 \cdot 1 = 1$. Then we have a shortest U-path $(h, (X \cup (1), \cdot))$ where h is the inclusion map of X in $X \cup (1)$. From this semigroup with unit we reach a group in two stages, using the methods in (c*) and (d*) in Exercise 1; this really provides a shortest (UK^*J^*)-path from $(X, \cdot)$ to the category Gp. By Lemma 7.6.2, the semigroup $(X, \cdot)$ would be isomorphic with a subsemigroup of a group iff this shortest path leads to a monomorphism, or injective map, of $(X, \cdot)$ in the group

$(W(X^* \cup X^{*\prime}), \circ)/E$ where X^* is the semigroup obtained from X by adjunction of unit element and E is the congruence defined in (d*). Hence if p, q are elements of X for which $(p, q) \in E$, then $p = q$ must be true. This implication gives rise to a whole sequence of implications, called the *Malčev conditions* for embeddability of the semigroup $(X, \cdot)$ in a group. We examine some typical ones:

(M1) $[p \cdot r = s, q \cdot r = s$ in $X]$ imply $[p = q]$; for we can pass from p to q through a sequence of words of the form p, (p, r, r'), (s, r'), (q, r, r'), q, showing that (p, q) must be in E, hence $p = q$.

(M2) $[a \cdot b = c \cdot d, p \cdot b = q \cdot d$, and $a \cdot m = c \cdot n]$ imply $[p \cdot m = q \cdot n]$: for we can show that $(p \cdot m, q \cdot n)$ is in E, since we can pass from $p \cdot m$ to $q \cdot n$ through the sequence of words $(p \cdot m)$, (p, m), (p, a', a, m), (p, a', c, n), (p, a', c, d, d', n), (p, a', a, b, d', n), (p, b, d', n), (q, d, d', n), (q, n), $(q \cdot n)$.

The main contention due to Malčev is that we could have such implications with the premise containing as many equalities as we please.

3. In Theorem 7.6.2 we gave proofs for a selection of typical cases of inclusion functors between various GF-algebras. Try to work out proofs for some of the remaining cases.

4. Use the Special Adjoint Functor theorem to prove the existence of a left-adjoint for the inclusion functor C: CSU → SU (of Lemma 7.6.3), using a method similar to that for the Stone–Cěch-compactification (under Lemma 7.6.4).

5. We constructed a shortest K-path for any object $(X, \leq)$ of M^* in Theorem 7.8.1. But if we assume K to be the functor from DS to the larger category M, this construction no longer works. It may not even give a least K-path: for consider the object $(X, \leq)$ of M illustrated in Figure 7.17; the construction of the lattice $(X\#, \leq)$ as in Theorem 7.8.1 gives the lattice that is also illustrated. Another object from DS is $(X', \leq')$, and a morphism g: $(X, \leq) \to (X', \leq')$ is described by setting $g(x) = 0'$ for $x = p, q, r$ or 0, while g takes a, b, c to a', b', c', respectively. This is a morphism of M (but not of M^*). There is no morphism g^*: $(X\#, \leq)$ to $(X', \leq)$ for which $K(g^*) \cdot p = g$; for any such g^* must take $p(p)$, $p(q)$, $p(r)$ to $0'$, and then $p(a) = \Sigma^*[p(p), p(q)]$ also to $0'$, similarly, all of the elements of $X\#$ to $0'$, and then $K(g^*) \cdot p(a) = 0' \neq g(a) = a'$.

6. The triads $(A°, F°, G°)$ and (A^*, F^*, G^*) that we associated with a given monad on B in Theorem 7.7.1 are due to Kleisli [18] and to Eilenberg and Moore [9]. In Appendix 2 we shall relate the second one to the definition of "algebraic categories" over a base category. We shall prove here the uniqueness of the functors K, L that we constructed in Theorem 7.7.1.

As a preparation for the proof we establish another useful result:

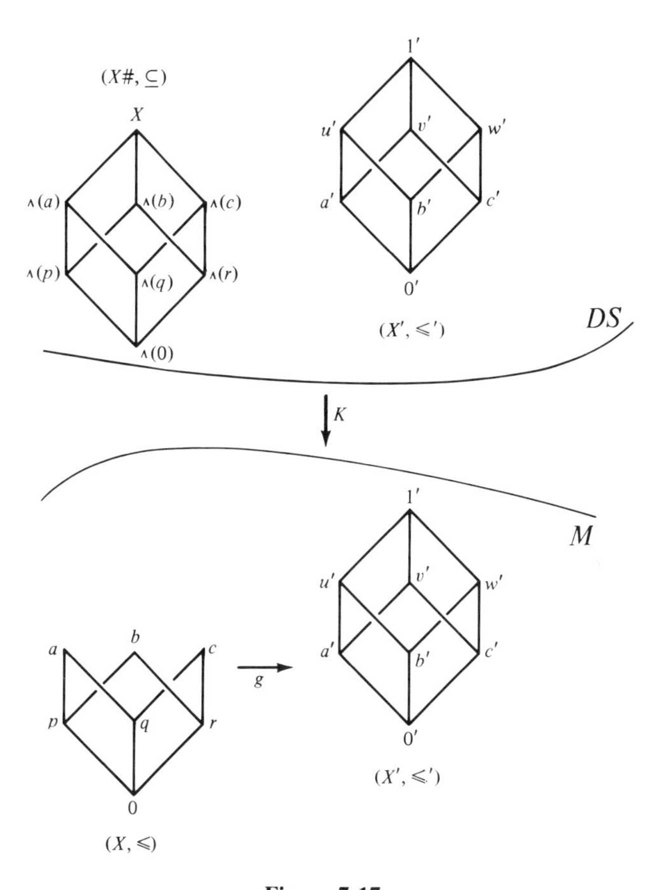

Figure 7.17

6(a). If (A, F, G) and (A', F', G') are two triads that determine the same monad (H, η, δ) on B, through the near-equivalences $(\eta, \nu; F, G)$ and $(\eta, \nu'; F', G')$, and if $(A, F, G) \leq (A', F', G')$, the functor $L: A \to A'$ for which $L \cdot F = F'$ and $G' \cdot L = G$ also satisfies $L\nu = \nu' L$ [that is, for any object A of A, $L(\nu(\mathrm{A})) = \nu'(L(\mathrm{A})]$. From Figure 7.18 it is clear that $G(\nu(\mathrm{A})) \cdot \eta(G(\mathrm{A})) = 1_{G(\mathrm{A})}$ and $\nu'(L(\mathrm{A}))$ is the unique morphism h in $\hom_{A'}(F'G'L(\mathrm{A}), L(\mathrm{A}))$ for which $G'(h) \cdot \eta(G(\mathrm{A})) = 1_{G(\mathrm{A})}$; but the earlier result asserts, by using $G = G' \cdot L$, that $L(\nu(\mathrm{A}))$ is also a morphism in $\hom_{A'}(LFG(\mathrm{A}), L(\mathrm{A})) = \hom_{A'}(F'G'L(\mathrm{A}), L(\mathrm{A}))$ for which $G'[L(\nu(\mathrm{A}))] \cdot \eta(G(\mathrm{A})) = 1_{G(\mathrm{A})}$; hence the uniqueness of the h asserted just now means that $L(\nu(\mathrm{A})) = \nu'(L(\mathrm{A}))$.

6(b). For $(A^\circ, F^\circ, G^\circ) \leq (A', F', G')$, when (A', F', G') is any triad determining the same monad (H, η, δ) on B as $(A^\circ, F^\circ, G^\circ)$, the associated functor K: $A^\circ \to A'$ is unique. We already defined one such K in the proof of the theorem, with $K[f: \mathrm{A} \to \mathrm{B}] = [\nu' F'(\mathrm{B}) \cdot F'(f) \cdot F'\eta(\mathrm{A}): F'(\mathrm{A}) \to F'(\mathrm{B})]$. Let us assume that for a K': $A^\circ \to A'$ we have again $K' \cdot F^\circ = F'$ and

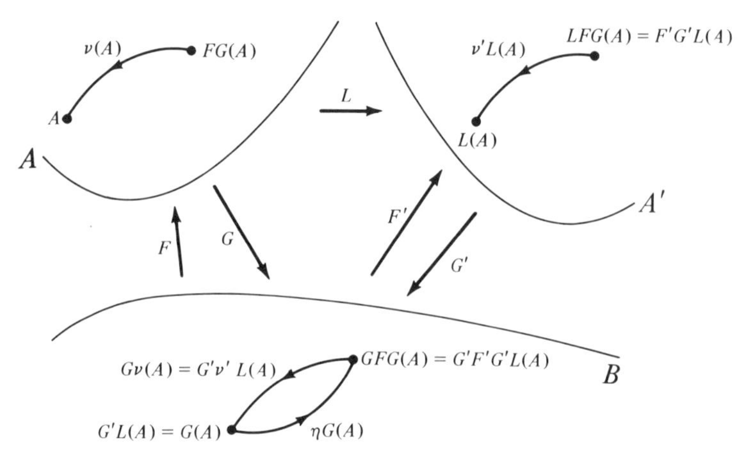

Figure 7.18

$G' \cdot K' = G^\circ$. For an object A of A°, since A is also an object of B, $K' \cdot F^\circ(\mathrm{A}) = F'(\mathrm{A})$ gives $K'(\mathrm{A}) = F'(\mathrm{A})$. If (A', F', G') determines the monad through the near-equivalence $(\eta, \nu'; f', g')$, we have the commutativity relations for the morphisms $K(f)$, $K'(f)$ both between $F'(\mathrm{A})$ and $F'(\mathrm{B})$ in A' of the form $K(f) \cdot \nu' F'(\mathrm{A}) = \nu' F'(\mathrm{B}) \cdot F'G'K(f)$, $K'(f) \cdot \nu' F'(\mathrm{A}) = \nu' F'(\mathrm{B}) \cdot F'G'K'(f)$. Since $G' \cdot K = G' \cdot K'$, the second parts of the two relations are equal, so we get $K(f) \cdot \nu' F'(\mathrm{A}) = K'(f) \cdot \nu' F'(\mathrm{A})$; but $\nu' F'(\mathrm{A})$ is a retraction, so an epimorphism; hence $K(f) = K'(f)$ also, proving that $K' = K$.

6(c). The functor $L\colon A' \to A^*$ giving the relation $(A', F', G') \le (A^*, F^*, G^*)$ is unique. We already saw that $L(\mathrm{A}') = (G'(\mathrm{A}'), G'\nu'(\mathrm{A}'))$ for any object A′ of A', and $L(f') = G'(f')$ for any morphism f' of A'. If we now assume that a functor $L'\colon A' \to A^*$ also satisfies the conditions $L' \cdot F' = F^*$ and $G^* \cdot L' = G'$, we see at once that for a morphism f' of A', if $L'(f') = f$, then $G'(f') = G^* \cdot L'(f') = G^*(f) = f$ (from the definition of G^*). Thus $L'(f') = G'(f') = L(f')$. For an object A′ of A', if $L'(\mathrm{A}') = (\mathrm{A}, a)$ in A^*, $G^* \cdot L' = G'$ gives $G^*(\mathrm{A}, a) = G'(\mathrm{A}')$ or $\mathrm{A} = G'(\mathrm{A}')$ (from the definition of G^* on objects of A^*). Using result 6(a) once with L and once with L', we get $\nu^*(L(\mathrm{A}')) = L(\nu'(\mathrm{A}'))$ and $\nu^*(L'(\mathrm{A}')) = L'(\nu'(\mathrm{A}'))$, but for the morphism $\nu'(\mathrm{A}')$ from A' we already saw that L and L' have the same effect; thus $L(\nu'(\mathrm{A}')) = L'(\nu'(\mathrm{A}'))$; hence $\nu^*(L(\mathrm{A}')) = \nu^*(L'(\mathrm{A}'))$, or $\nu^*(G'(\mathrm{A}'), G'\nu'(\mathrm{A}')) = \nu^*(G'(\mathrm{A}'), a)$; but for an object (A, a) of A^* it is easy to see that $\nu^*(\mathrm{A}, a) = a$; thus $a = G'\nu'(\mathrm{A}')$ and $L'(\mathrm{A}') = L(\mathrm{A}')$ for any object A′ of A' too. Thus $L' = L$.

APPENDIX ONE

SEMIUNIFORM, BITOPOLOGICAL, AND PREORDERED ALGEBRAS

As a typical example of this form of structure we have the additive Abelian group $(R, +)$ of real numbers with the semiuniformity on it defined by the semimetric d^* given by $d^*(x, y) = \max[0, y - x]$; although the binary operation $(a, b) \to (a + b)$ would be covariant in both terms, the operation $(a, b) \to (a - b)$ has to be considered covariant in the first and contravariant in the second term. We shall see how these variances can be worked into the definition of operations. Before that, we collect some features common to the categories *SU, BT, PO, S*, etc.

Definition A.1.1 A functor $N: G \to G$ is called a *conjugation* on the category G if $N \circ N = N$; that is, $N(N(\mathrm{A})) = \mathrm{A}$ and $N(N(f)) = f$ for any object A and any morphism f of G.

Any category G has a trivial conjugation, namely 1_G. For S we always take this 1_S as its basic conjugation. In *SU* we have a conjugation N taking $f: (X, U) \to (Y, V)$ to $f: (X, U^r) \to (Y, V^r)$; in *BT* we have a conjugation taking $f: (X, T, T') \to (Y, S, S')$ to $f: (X, T', T) \to (Y, S', S)$; and in *PO* we have a conjugation taking $f: (X, \leq) \to (Y, \leq)$ to $f: (X, \geq) \to (Y, \geq)$. These we take as the standard conjugations in these three categories.

Definition A.1.2 A triple (G, N, B) is a *fine category* if G is a category closed for products of sets of its objects, with a conjugation N and a faithful product-preserving functor $B: G \to S$ that also preserves conjugation, so that $B(\mathrm{A}) = B(N(\mathrm{A}))$ and $B(f) = B(N(f))$ for A in $O(G)$, f in $M(G)$.

Definition A.1.3 A functor $F: G \to H$ is called a *fine functor* from the fine category (G, N, B) to the fine category (H, N', B') if $B' \cdot F = B$, $F \cdot N = N' \cdot F$, and F is a faithful functor preserving products.

When (G, N, B) is a fine category, it is clear that $B: G \to S$ is a fine functor from (G, N, B) to $(S, 1_S, 1_S)$. For convenience we generally denote the conjugation by $(\ )'$ and the base functor by $|\ |$ in all fine categories. Thus $(A)'$ and $(f)'$ denote the conjugates of A and f, while $|A|$ and $|f|$ denote the set bases of A and f (the effect of the base functor to S); and we talk simply of the fine category G (instead of the fine category $(G, (\ )', |\ |)$).

In defining a typical algebraic operation on an object in a fine category G, we require a name θ for the operation, the integer n denoting its arity, and a symbol v for its variance whenever $n \geq 1$. For $n = 0$, or a nullary operation, we just have the name θ for the operation. We then consider a family F that is a disjoint union of F_0, F_1, the subfamilies of the nullary and the nonnullary operations, respectively. Each element of F_0 is just a symbol (θ) for a nullary operation, whereas an element of F_1 is a triple (θ, n, v) with n an integer ≥ 1 and v a mapping of the sequence $(1, \ldots, n)$ in the two-element set $[+1, -1]$. We say that this operation is covariant/contravariant at the place m $(1 \leq m \leq n)$ according as $v(m) = +1$ or -1.

Definition A.1.4 An object A of the fine category G is said to be closed for (i) a θ of F_0 if $|A|$ is a nonnull set and $\theta(A)$ [or $\theta^A(\varnothing)$] is a uniquely assigned element from $|A|$;

(ii) a (θ, n, v) of F_1 if there is assigned a morphism $\theta^A: P[A_1, \ldots, A_n] \to A$ of $M(G)$, where P denotes a product object in G of the A_i, and $A_i = A$ or $(A)'$ according as $v(i) = +1$ or -1.

Definition A.1.5 An object A of G is called an F-algebra over G (or a *GF-algebra*) if A is closed for each operation from F.

An operation $\theta^* = (\theta, n, v)$ from F_1 determines an associated functor $P(\theta^*): G \to G$ when we set $P(\theta^*)(A) = P[A_1, \ldots, A_n]$ with $A_i = A$ or $(A)'$ according as $v(i) = +1$ or -1, and we set $P(\theta^*)[f: A \to B]$ equal to the morphism $P(f_1, \ldots, f_n): P(A_1, \ldots, A_n) \to P(B_1, \ldots, B_n)$ defined by means of the morphisms $f_i: A_i \to B_i$, where $f_i = f$ or $f_i = f': (A)' \to (B)'$ $(=(f)')$ according as $v(i) = +1$ or -1. We say that a mapping $f: |A| \to |B|$ is an *F-homomorphism* of the F-algebra A in the F-algebra B (both over G) if for each θ from F_0, $|f|(\theta(A)) = \theta(B)$, and for each θ^* from F_1, $f \cdot \theta^A = \theta^B \cdot P(\theta^*)(f)$.

We observe that when A is an F-algebra over G, we can so define the operations of F on $(A)'$ and on $|A|$ that these become F-algebras over G

and over S, respectively, with $\theta(A') = \theta(A) = \theta(|A|)$ for any θ of F_0, and for a $\theta^* = (\theta, n, v)$ of F_1, $\theta^{A'} = [\theta^A]'$, where we take $P[A_1', ..., A_n']$ to be the same as $(P[A_1, ..., A_n])'$, and $\theta^{|A|} = |\theta(A)|$.

It is clear then that we have a category GFA whose objects are the GF-algebras and whose morphisms are the GF-morphisms between such algebras; this category is a fine category with obvious definitions for ( )', | |; and a fine functor K: $G \to H$ gives rise to a functor, also denoted by K, from GFA to HFA that is also a fine functor.

When we take S for G, we usually denote SFA by just FA; this has as objects the usual F-algebras, and for these the variance v in a θ^* of F_1 is really ineffective, since the conjugation is the identity here.

We next see how to define FD-algebras over G, which for $G = S$ give our older FD-algebras.

Given a set D of pairs from the free F-algebra $P(F, A)$ over some set A, we have already defined what is meant by an FD-algebra (X, F). Given the fine category G, we now define an object X of G to be an *FD-algebra over G* if, first, it is an F-algebra over G, and further, $|X|$ is an FD-algebra (over S). With these FD-algebras over G as objects we get a full subcategory $GFDA$ of GFA. Again, when T: $G \to H$ is a fine functor between two fine categories (G, B), (H, B'), T can also be treated as a conjugation-preserving faithful functor T: $GFDA \to HFDA$.

Theorem A.1.1 *(a) All the categories CSU, SU, BT, PO, and S are fine categories, under a natural definition of the base functor for each. There are fine functors C: CSU →SU, B: SU → BT, P: BT → PO, and S: PO → S.*

(b) The preceding four functors and the similarly named ones between F-algebras or FD-algebras over the G-categories involved give 12 conjugation-preserving and faithful functors, each of which has a left-adjoint.

PROOF. (a) The base functor from CSU, SU, BT, or PO is just the functor that forgets the extra structure: the semiuniformity, bitopology, or preorder. For S, 1_S is the base functor.

We have already defined the functors B, C in Lemma 7.6.3. The functor P from BT to PO is the composite of the two functors L: $BT \to T$ and $\mathring{\leq}$: $T \to PO$ as defined in that lemma; that is, for an (X, T, T'), $P(X, T, T') = (X, \leq)$ where $\leq$ is the same as $\leq(T) = \geq(T')$; while S: $PO \to S$ is the base functor for PO.

To prove that these are fine functors, we check one, B: $SU \to BT$; the others can all be treated similarly. In SU, a typical object is $(X, U, (J, \leq), c)$, or (X, U), and a typical morphism is a uniform map f: $(X, U) \to (Y, V)$. In BT an object is of the form (X, T, T') and a morphism of the form f: $(X, T, T') \to (Y, S, S')$; and $B(X, U) = (X, T(U), T(U^r))$. Both the base functor of SU and the composite of B followed by the base functor of BT have the same effect on objects and morphisms of SU. B is a faithful functor, since the two base functors are faithful. From the definitions it is clear that B preserves conjugation. Also, from the way products are defined in SU and BT (in both cases over a set base that is a direct product of bases of the single terms), it is not difficult to check that B takes products to products. These prove that B is a fine functor. The others are proved similarly.

(b) The 12 conjugation-preserving faithful functors have already been defined. There are three named C, three named B, three named P, and three named S. We therefore treat them in triples.

The Three Functors C. For the functor C: $CSU \to SU$ we know from Theorem 1.6.1 that each object (X, U) of SU has a shortest C-path $[h, (X\#, U\#)]$, and so C has a left-adjoint. We recall from Exercise 24 at the end of Chapter 1 giving a proof of Theorem 1.6.1 the way $(X\#, U\#)$ is obtained from $(X, U, (J, \leq), c)$. We first constructed a complete semiuniform space $(X', U', (J, \leq), c)$ with X' as the set of Cauchy $(J, \leq)$-sequences $[x']$ from (X, U); then defined $[(x', y') \in U'(j)]$ iff [there is a nonnull initial subset $B(j)$ of $(J, \leq)$ such that $(x'(k), y'(k))$ is in $U(j)$ for each k of $B(j)$]. Then using the equivalence E' on X' defined by U', we formed the quotient space $(X'/E', U'/E')$ and this was $(X\#, U\#)$; for $(x\#, y\#)$ to be in $U\#(j)$ the condition was that for any choice of an element x' of X' in the E'-class $x\#$ there must be an element y' in the E'-class $y\#$ such that (x', y') is in $U'(j)$. Also, $h(x) = [(x^c)]^E$.

To get the left-adjoint then of C: $CSUF \to SUF$, we show that there is a shortest C-path $[h, (X\#, U\#, F)]$ for each object (X, U, F) of SUF, with $(X\#, U\#)$ defined from (X, U) as above. So we have to extend the F-operations to $X\#$ from the way they are defined in X. Note that we are thinking now of an object (X, U, F) as an F-algebra (X, F) and a semiuniform space (X, U) with the further property that the set map θ^X: $P[X_1, \ldots, X_n] \to X$, where each $X_n = X$, leads to a uniform map $\theta^{(X,U,F)}$: $P[(X, U_1), (X, U_2), \ldots, (X, U_n)] \to (X, U)$ where $U_i = U$ or U^r according as $v(i) = +1$ or -1. This has to be done for each (θ, n, v) of F_1; for a v from F_0, $v^{(X,U)}(\varnothing) = v^X(\varnothing)$ by assumption. Nothing need be proved for these nullary operators. We shall do this extension of the typical nonnullary operator (θ, n, v) in two stages through (X', U') to $(X\#, U\#)$. Given an n-tuple of elements $(x_1', \ldots, x_n')$ from X', we associate an element $x' = \theta^{X'}(x_1', \ldots, x_n')$ of X' as follows: $x'(j) = \theta^X(x_1'(j), x_2'(j), \ldots, x_n'(j))$ for each j of J. It is easy to verify that x' is a Cauchy $(J, \leq)$-sequence from (X, U) when each of the x_i' $(i = 1, \ldots, n)$ is so. Also, it can be verified that given a j of J, a k of J can be found such that for n pairs $(x_1', y_1'), \ldots, (x_n', y_n')$ from X' such that $[(x_i', y_i') \in U_1'(k)$ for $i = 1, \ldots, n$, where $U_1'(k)$ is either $U'(k)$ or $(U^r)'(k) = (U'(k))^r$ according as $v(i) = +1$ or -1], we have that $[\theta^{X'}(x_1', \ldots, x_n'), \theta^{X'}(y_1', \ldots, y_n')]$ is in $U'(j)$; this follows from the assumption that (X, U, F) is a semiuniform F-algebra. Thus $\theta^{X'}$ is a morphism with the right type of variance to provide a semiuniform F-algebra (X', U', F) based on (X', U'). The extension of the operation (θ, n, v) to $X\#$ is then not difficult. For, given an n-tuple $(x_1\#, \ldots, x_n\#)$ of elements from $X\#$, if we choose pairs of elements (x_1', y_1') from $x_1\#$, (x_2', y_2') from $x_2\#$, ..., (x_n', y_n') from $x_n\#$, then our earlier remarks show that the elements $\theta^{X'}(x_1', \ldots, x_n')$ and $\theta^{X'}(y_1', \ldots, y_n')$ are E'-equivalent. So choosing arbitrarily one element $(x_1', \ldots, x_n')$ from the sets $(x_1\#, \ldots, x_n\#)$ in order and then forming the E'-class of $\theta^{X'}(x_1', \ldots, x_n')$ gives a unique element of $X\#$, irrespective of the choice of the x_i'; this we then denote by $\theta^{X\#}(x_1\#, \ldots, x_n\#)$.

To verify now that this operation defined in $X\#$ has the proper variance, assume that a set of n pairs of elements $(x_1\#, y_1\#), \ldots, (x_n\#, y_n\#)$ is given. We have to show that for a given j of J there is a k of J such that, if $(x_i\#, y_i\#)$ is in $U_i\#(k)$ for each i $(i = 1, 2, \ldots, n)$, then $[\theta^{X\#}(x_1\#, \ldots, x_n\#), \theta^{X\#}(y_1\#, \ldots, y_n\#)]$ belongs to $U\#(j)$ where $U_i\#(k)$ is either $U\#(k)$ or $U\#(k)^r$, according as $v(i)$ is $+1$ or -1. For the conclusion to be true, given an element of the E'-class $\theta^{X\#}(x_1\#, \ldots, x_n\#)$, say x', we have to find an element y' of the E'-class $\theta^{X\#}(y_1\#, \ldots, y_n\#)$ such that (x', y') is in $U'(j)$. Starting with j and cj, we choose a k (for cj) such that $[(x_i', y_i') \in U_i'(k)$ for $i = 1, \ldots, n]$ would imply that $[\theta^{X'}(x_1', \ldots, x_n'), \theta^{X'}(y_1', \ldots, y_n')]$ would be in $U'(cj)$. Then for the given x' of $\theta^{X\#}(x_1\#, \ldots, x_n\#)$, if we make any choice of elements x_i' from the classes $x_i\#$, $(x', \theta^{X'}(x_1', \ldots, x_n'))$ belongs to E' and hence to $U'(cj)$; for each x_i' from $x_i\#$ we can find a y_i' from $y_i\#$ such that (x_i', y_i') is in $U_i'(k)$ if we are given that $(x_i\#, y_i\#)$ is in $U_i\#(k)$ (for each i). Then $[\theta^{X'}(x_1', \ldots, x_n'), \theta^{X'}(y_1', \ldots, y_n')]$ would be in $U'(cj)$; hence we have $(x', \theta^{X'}(y_1', \ldots, y_n'))$ is in $U'(j)$ for an element $(\theta^{X'}(y_1, \ldots, y_n))$ from $\theta^{X\#}(y_1\#, \ldots, y_n\#)$. This proves the necessary variance property to make $(X\#, U\#, F)$ a semiuniform F-algebra. So we have defined $(X\#, U\#, F)$.

To show then that $(h, (X\#, U\#, F))$ is a shortest C-path from the object (X, U, F) of SUF (to the category $CSUF$), first we note that h is an F-morphism from (X, F) to $(X\#, F)$; this follows from the way the F-operations were defined in $X\#$ and the way that h is defined. If there were a morphism g: $(X, U, F) \to (Y, V, F)$ in USF, where (Y, V, F) also belongs to $CUSF$ [so it is a complete Hausdorff semiuniform space (Y, V)], we would have to show that the semiuniform map g^*: $(X\#, U\#) \to (Y, V)$ associated with this g (such that $g^* \cdot h = g$) is also an F-homomorphism. It suffices to show that for a typical (θ, n, v) of F_1 and n-tuple of elements $(x_1\#, \ldots, x_n\#)$ from $X\#$, if $g^*(x_1\#) = y_i$ for $i = 1, \ldots, n$, then $g^*[\theta^{X\#}(x_1\#, \ldots, x_n\#)] = \theta^Y(y_1, \ldots, y_n)$. Recall that $g^*(x_1\#) = y_i$ implies that if we choose an x_i' (of X') from $x_i\#$, then the Cauchy $(J, \leq)$-sequence $[g(x_i'(j)): j \text{ in } J)]$ of (Y, V) converges (under $T(V^*)$) to y_i. If we now choose one x_i' from each $x_i\#$, then $\theta^{X'}(x_1', \ldots, x_n')$ is an element in the E'-class containing $\theta^{X\#}(x_1\#, \ldots, x_n\#)$, so $g^*[\theta^{X\#}(x_1\#, \ldots, x_n\#)]$ can be taken to be the (unique) limit in $(Y, T(V^*))$ of the Cauchy sequence $[g(\theta^{X'}(x_1', \ldots, x_n'))] = [g(\theta^X(x_1'(j), \ldots, x_n'(j))): j \text{ in } J] = [\theta^Y(g(x_1'(j)), g(x_2'(j)), \ldots, g(x_n'(j)): j \text{ in } J]$ since g is an F-homomorphism. Since (Y, V, F) is a semiuniform F-algebra, in the associated Hausdorff space $(Y, T(V^*))$ the last-named sequence must converge to $\theta^Y(y_1, \ldots, y_n)$, since the sequence $[g(x_i'(j)); j \text{ in } J]$ converges to y_i for each i. This shows that g^* is an F-homomorphism also. So we have a shortest C-path for each object (X, U, F) from SUF (to $CSUF$); this C has a left-adjoint.

For the third, C: $CSUFDA \to SUFDA$, the only step beyond what we did in the last case would be to show that when we start with an object (X, U, F) from $SUFD$ and go to the completion $(X\#, U\#, F)$, this F-algebra is also an FD-algebra. For this we have to prove that given a set map $g: A \to X\#$, if g^* is the associated F-homomorphism of $(P(A, F), F)$ in $(X\#, F)$, then $g^*(p) = g^*(q)$ whenever (p, q) is from D. Let us define a map $e: X\# \to X'$ such that $e(x\#)$ is an element of the E'-class $x\#$; and denote by $p(j)$ the map of X' in X that chooses for each x' of X' its jth element $x'(j)$. We have then the following functions (or set maps) from A: g as given to $X\#$; $e \cdot g$ to X'; and for each j of J a $p(j) \cdot e \cdot g$ to X. Each of these has an associated F-homomorphism from $(P(A, F), F)$ to (in order) $(X\#, F)$, (X', F), and (X, F). We use the same notation in all the cases: the associate of an h is an h^*. Then, since (X, F) is known to be an FD-algebra, for each j of J we have $(p(j) \cdot e \cdot g)^*(p) = (p(j) \cdot e \cdot g)^*(q)$. Now $(p(j) \cdot e \cdot g)^*(p)$ equals $p(j)[(e \cdot g)^*(p)]$ from the way the operators are defined in X'. Hence, $(e \cdot g)^*(p)$ and $(e \cdot g)^*(q)$ have the same jth component for each j; so they must be equal as elements of X'. But then $(e \cdot g)^*(p)$, $(e \cdot g)^*(q)$ are surely contained in the E'-classes $g^*(p)$ and $g^*(q)$. Hence these two E'-classes, having a common element, must be identical, proving what we wanted, namely, $g^*(p) = g^*(q)$.

The Functors B. In Lemma 7.6.3 we proved that B: $SU \to BT$ has a left-adjoint by showing that for any object (X, T, T') of BT there is a shortest B-path $(1_X, (X, U^*))$ where U^* is the finest among the semiuniformities U on X for which the mappings 1_X: $(X, T) \to (X, T(U))$ and 1_X: $(X, T') \to (X, T(U^r))$ are both continuous.

Next is B: $SUFA \to BTFA$. For an object $(X, T, T'; F)$ that is in BTF there are semiuniformities U on X such that (i) the F-algebra (X, F) has the operators of the right variances to make it a semiuniform F-algebra; (ii) the maps 1_X: $(X, T) \to (X, T(U))$ and 1_X: $(X, T') \to (X, T(U^r))$ are both continuous; for the coarsest semiuniformity on X surely satisfies these conditions. If we take the lattice product U^* of these semiuniformities on X, it can also be seen to satisfy these conditions. Thus (X, U^*, F) is an object of *SUFA* and it is easy to see that $(1_X, (X, U^*, F))$ gives a shortest B-path at $(X, T, T'; F)$. So this B also has a left-adjoint.

For B: SUFDA $\rightarrow$ BTFDA, the same construction as in the last paragraph leads from a bitopological FD-algebra $(X, T, T'; F)$ to a semiuniform FD-algebra (X, U^*, F); surely then $(1_X, (X, U^*, F))$ gives a shortest B-path at the object $(X, T, T'; F)$ of *BTFDA*. Hence this B also has a left-adjoint.

The Functors P. Given P: $BT \rightarrow PO$, to any object $(X, \leq)$ from *PO* we associate the object (X, T, T') of *BT*, where T = [the family of all initial sets of $(X, \leq)$] and T' = [the family of all final sets of $(X, \leq)$]. It is not hard to see that $(1_X, (X, T, T'))$ gives a shortest P-path at $(X, \leq)$.

For P: $BTFA \rightarrow POFA$, again for an object $(X, \leq; F)$ from *POFA*, which is a preordered F-algebra, we consider the pair of topologies T, T', as earlier, and check that $(X, T, T'; F)$ is a bitopological F-algebra. This involves proving, given a typical (θ, n, v) of F_1, that the set mapping $\theta^X[X^n] \rightarrow X$ has the proper variance when the same mapping is to be considered a morphism of $P[(X_i, T_i, T_i')] \rightarrow (X, T, T')$, where $(X_i, T_i, T_i') = (X, T, T')$ or (X, T', T) according as $v(i) = +1$ or -1. We already have the information that this set map has the right variances to make it a morphism from $P[(X_i, \leq_i)] \rightarrow (X, \leq)$ where $(X_i, \leq_i) = (X, \leq)$ or $(X, \geq)$ according as $v(i) = +1$ or -1. That which has to be proved here is, if $(x_1, y_1), \ldots, (x_n, y_n)$ is a set of n pairs of elements from X, [given a T-open set G of X containing $\theta(x_1, \ldots, x_n)$, there exist for each i, a T_i-open set G_i containing x_i (where $T_i = T$ or T', according as $v(i) = +1$ or -1), such that y_i is in G_i for each i, implies $\theta(y_1, \ldots, y_n)$ also belongs to G; and symmetrically, given a T'-open set G' containing $\theta(x_1, \ldots, x_n)$, there are T_i'-open sets G_i' containing the x_i (with $T_i' = T'$ or T according as $v(i) = +1$ or -1) such that $\theta(y_1, \ldots, y_n)$ belongs to G' if each y_i is from the corresponding G_i']. These can be proved: for if G is a T-open set containing $x = \theta(x_1, \ldots, x_n)$, it must be an initial set containing x. For the G_i we choose the sets $L_i(x_i)$ where $L_i(x_i)$ = [set of $y \leq x_i$ or the set of $y \geq x_i$ according as $v(i) = +1$ or -1]; these are then such that y_i in L_i for each i would imply that $y = \theta(y_1, \ldots, y_n)$ is $\leq x$ and hence in G, since $(X, \leq; F)$ is a preordered F-algebra. Similarly, given the G' that is a final subset containing x, we choose the G_i' to be the $L_i'(x_i)$ = [the set of $y \geq x_i$ or the set of $y \leq x_i$ according as $v(i)$ is $+1$ or -1]. Again, if each y_i is in L_i' it would follow that $y \geq x$, and so belongs to G'. Thus for the object $(X, \leq; F)$ from *POF* we have a shortest P-path $(1_X, (X, T, T'; F))$. So this P has a left-adjoint.

For P: $BTFDA \rightarrow POFDA$ this same construction works; when the original F-algebra $(X, \leq; F)$ is an FD-algebra the same is true of the $(X, T, T'; F)$.

The Functors S. These are very simple. Given S: $PO \rightarrow S$ to any object

X from S we associate $(X, =)$ from PO. Then $(1_X, (X, =))$ is easily seen to be a shortest S-path at X. Similarly, if we start with an (X, F) that is an F-algebra or an FD-algebra, then $(1_X, (X, =; F))$ again gives a shortest S-path for the object (X, F) of FA or FDA. So all these S have left-adjoints. □

So far we have considered cases of types of GA-categories and functors between pairs of them when the A-structure is kept fixed and only the G-structure is changed. We now look for the other type, when the G-structure is kept fixed and the A-structure varies through the types S, FA, FDA, F^*A, and F^*D^*A, where F^* contains F and D^* contains D, as in (7.6.1). The case when the G-structure is S is essentially covered in Lemma 7.6.1. We consider the other types one by one, based on a form of proof common to all of them.

To provide this common form of proof we introduce some extra notions and establish a basic lemma.

Definition A.1.6 Given a functor $H: D \to B$ between categories D, B, we say that:

(i) *H raises factorizations* if, given a morphism f of D and a factorization $H(f) = g_1 \cdot g_2$ in B, there is a factorization $f = f_1 \cdot f_2$ in D such that $H(f_i) = g_i$ for $i = 1, 2$;

(ii) *H raises codomains* if, given a D of $O(D)$ and a $g: H(\mathrm{D}) \to \mathrm{B}'$ in B, there is a D' in $O(D)$ and an $f: \mathrm{D} \to \mathrm{D}'$ in D such that $H(\mathrm{D}') = \mathrm{B}'$ and $H(f) = g$;

(iii) *H is latticiel* if the following two conditions are true:

(a) for each B of $O(B)$, $H^{-1}(\mathrm{B}) = [\mathrm{D} \text{ in } O(D): H(\mathrm{D}) = \mathrm{B}]$ is a set that is a complete lattice under a preorder $\leq(H)$ on it defined by $\mathrm{D} \leq (H)\, \mathrm{D}'$ for D, D' of $H^{-1}(\mathrm{B})$ iff there is a morphism $h: \mathrm{D} \to \mathrm{D}'$ in D such that $H(h) = 1_{\mathrm{B}}$;

(b) for each g of $M(B)$ and any set of morphisms $f_i: \mathrm{D} \to \mathrm{D}_i$ in D all having the same domain D and with $H(f_i) = g$ for each i, there is a morphism $\Pi(f_i)$ in D from D to the lattice product $\Pi(\mathrm{D}_i)$ in the lattice $[H^{-1}(H(\mathrm{D})), \leq(H)]$.

Lemma A.1.1 *Given subcategories C, A of the categories D, B, with the inclusion functors $I_1: C \to D$ and $I_2: A \to B$, suppose that there is a functor $H: D \to B$ with the following properties: H is faithful, raises factorizations and codomains, is latticiel, and further [f of $M(D)$ belongs to $M(C)$] iff [$H(f)$ belongs to $M(A)$]. Then for an object D of D, [$H(\mathrm{D})$ has a shortest I_2-path (g, A)] implies that [D has a shortest I_1-path (g^*, C^*) such that $H(g^*) = g$, $H(C^*) = \mathrm{A}$]; we then say that H raises the I_2-path (g, A) to the I_1-path (g^*, C^*).*

As a corollary we have, under the same assumptions on H, that I_1 has a left-adjoint provided that I_2 has a left-adjoint.

PROOF. (See Figure A.1.) Starting with D of $O(D)$, let us assume that (g, A) is a shortest I_2-path at H(D) = B of $O(B)$. Since H raises codomains, there is a g_i: D → D_i in $M(D)$ such that $H(g_i) = g$, $H(D_i)$ = A; since A is in A, $H(1_{D_i}) = 1_A$ is in A and so 1_{D_i} and D_i are in C. We then write C_i for D_i, and have a morphism g_i: D → C_i with $H(g_i) = g$. Consider the subset of $H^{-1}(g)$ consisting of all the g_i with domain D, codomain in C, and $H(g_i) = g$; since H is latticiel, there is a morphism $\Pi(g_i)$: D → $\Pi(C_i)$, with $H(\Pi(g_i)) = g$, hence $H(\Pi(C_i))$ = A in A. As before, it follows that $\Pi(C_i)$ must be an object of C. So $[\Pi(g_i), \Pi(C_i)]$ is an I_1-path at D; we claim that it is a shortest I_1-path, too. For let (h, C′) be any I_1-path at D. We have to show that there is a unique t^*: $\Pi(C_i)$ → C′ in C such that $t^* \cdot \Pi(g_i) = h$. We know that [$H(h)$, H(C′)] is an I_2-path at B = H(D); since we assumed that (g, A) was a shortest I_2-path at B, there is a unique morphism k in A from A to H(C′) such that $k \cdot g = H(h)$. Since H raises fac-

Figure A.1

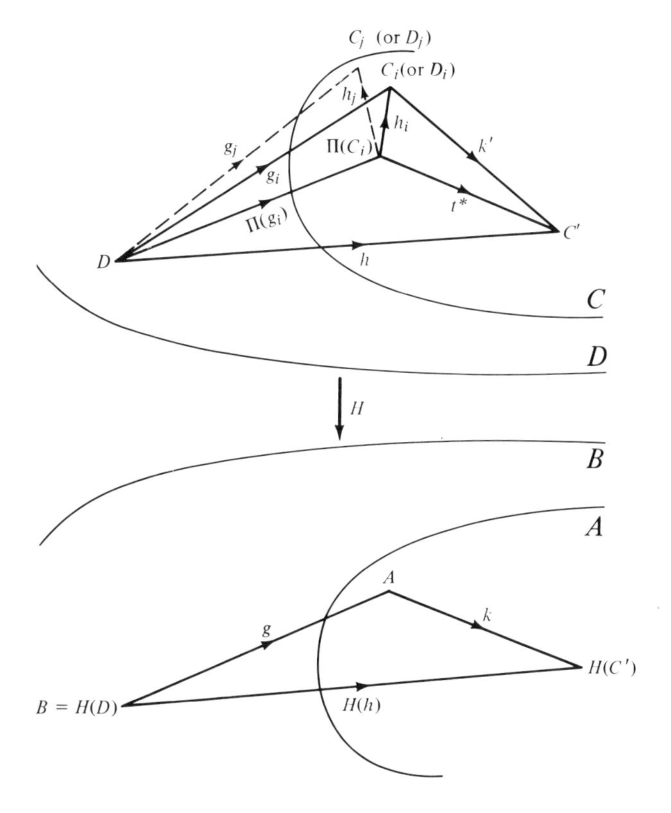

torizations, there is a corresponding factorization $h = k' \cdot g'$ of h with $H(k') = k$ in A, $H(g') = g$. From $H(k') = k$ it follows that k' is in C, so the codomain of g' equal to the domain of k' is an object of C. Hence g' must be one of the g_i; and then if C_i is the codomain of this $g' = g_i$, since $\Pi(C_i) \leq (H)\ C_i$, there is a morphism h_i: $\Pi(C_i) \to C_i$ such that $H(h_i) = 1_A$; hence $H(h_i \cdot \Pi(g_i)) = 1_A \cdot g = g = H(g_i)$. Since H is faithful, it follows that $h_i \cdot \Pi(g_i) = g_i$, so that $h = k' \cdot g_i = k' \cdot h_i \cdot \Pi(g_i)$. Hence we have a $t^* = k' \cdot h_i$ for which $h = t^* \cdot \Pi(g_i)$. The uniqueness of such a t^* follows then from the uniqueness of k and the fact that H is faithful. This proves that (g^*, C^*) is a shortest I_1-path at D if $g^* = \Pi(g_i)$ and $C^* = \Pi(C_i)$.

Since each object of D would have a shortest I_2-path if each object of B had a shortest I_1-path, the last statement follows from the part just proved, by Theorem 7.2.1. □

Now, using Lemma A.1.1 we can tackle the cases of GA-categories with G fixed and different A's and functors between these.

Theorem A.1.2 *(i) When G represents any one of the categories CSU, SU, BT, PO, or S, there are inclusion functors C^*: $GF^*D^*A \to GF^*A$, C': $GFDA \to GFA$, G^*: $GF^*D^*A \to GFDA$, G': $GF^*A \to GFA$, and H': $GFA \to G$. All 25 of these functors have left-adjoints.*

(ii) The four functors named in Theorem A.1.1, part (a), give rise to functors with the same symbols between F^*-algebras or F^*D^*-algebras over the G-categories involved [just as described in part (b) for F- and FD-algebras over the G-categories]. Thus there are 20 functors of the type given in Theorem A.1.1 among these 25 categories. All compositions of these 45 functors that are definable have left-adjoints.

PROOF. Part (ii) follows by using Theorem 7.4.1. For part (i) we indicate the proof for one typical example for each set of five cases with a fixed G. The proof is similar for the other four cases.

1. For $G = CSU$, let us choose the functor C': $CSUFDA \to CSUFA$. We use Lemma A.1.1 with $CSUFA$ and FA as the categories D and B, the forgetful functor from $CSUFA$ to FA as H, $CSUFDA$ and FDA as the categories C and A; then C' can be taken as I_1 and C': $FDA \to FA$ as I_2. In Lemma 7.6.1 we named this C' as C and showed that it had a left-adjoint. Thus I_2 has a left-adjoint. So if we verify that H has all the properties mentioned in Lemma A.1.1, then I_1 (or C') would also have a left-adjoint. We now check those properties for H; it is easy to see that H is faithful. If f: $(X, U; F) \to (Y, V; F)$ is a morphism in $CSUFA$ and $H(f) = f$: $(X, F) \to (Y, F)$ factorizes as $f = g_1 \cdot g_2$, g_2: $(X, F) \to (Z, F)$, g_1: $(Z, F) \to (Y, F)$, we can define for the F-algebra (Z, F) semiuniformities

W such that $(Z, W; F)$ is in $SUFA$, and g_1 is a uniform map of (Z, W) in (Y, V). In fact, there is a coarsest such W, and under it (Z, W) is a complete semiuniform space, so that $(Z, W; F)$ is in $CSUFA$; then g_2 is a uniform map of (X, U) in (Z, W). That is, H raises the factorization $f = g_1 \cdot g_2$ in FA to a factorization $f = g_1 \cdot g_2$ in $CSUFA$.

Given an object $(X, U; F)$ of $CSUFA$ and a morphism $g: (X, F) \to (Y, F)$ in FA, we can assign to Y its coarsest semiuniformity [with a single $V(j) = Y \times Y$] and get an object $(Y, V; F)$ of $CSUFA$ with $g: (X, U; F) \to (Y, V; F)$ a morphism of $CSUFA$. This means that H raises codomains.

For an object (X, F) of FA the possible semiuniformities U on X that make $(X, U; F)$ an object of $CSUFA$ are seen to be a complete lattice under the order $\leq(H)$, which is now just the relation of "being finer than." And given a set of morphisms $[f_i: (X, U; F) \to (Y, V_i; F)]$ from a common domain in $CSUFA$ for which $H(f_i): (X, F) \to (Y, F)$ is a unique morphism f of FA, it is clear that this f is also a uniform map of (X, U) in $(Y, \Pi(V_i))$, and that the space $(Y, \Pi(V_i))$ is complete, like the spaces $(Y, V_i; F)$. Thus H has been shown to be latticiel.

Finally, given a morphism $f: (X, U; F) \to (Y, V; F)$ in $D = CSUFA$, this f belongs to $C = CSUFDA$ iff (X, F) and (Y, F) are both FD-algebras; and this again is true iff $H(f) = f: (X, F) \to (Y, F)$ belongs not only to FA but also to FDA. Hence we have verified all parts of the hypotheses on H in Lemma A.1.1 to ensure that I_1, or C', should have a left-adjoint.

2. For $G = SU$, let us consider $C^*: SUF^*D^*A \to SUF^*A$. We are using Lemma A.1.1, setting $D = SUF^*A$ and $B = F^*A$, with H the forgetful functor from SUF^*A to F^*A. The subcategories SUF^*D^*A and F^*D^*A are taken as C and A; the inclusion functors I_1, I_2 are formed by the C^*'s. Since Lemma 7.6.1 gives a left-adjoint for I_2, we have only to check that this H has the requisite properties for applying Lemma A.1.1. This verification is almost a repetition of the similar one for the last case.

3. For $G = BT$, let us look at $G': BTF^*A \to BTFA$. We taken F^*A, FA, BTF^*A, and $BTFA$ as the categories A, B, C, and D, and H as the forgetful functor from $BTFA$ to FA. The inclusion functors I_1, I_2 are the two G''s; and since Lemma 7.6.1 again gives a left-adjoint for G' or I_2: $F^*A \to FA$, we have only to check that H has the required properties to apply Lemma A.1.1. Clearly H is faithful. It raises factorizations; for if $f: (X, T, T'; F) \to (Y, S, S'; F)$ is a morphism in $BTFA$ and $H(f)$: $(X, F) \to (Y, F)$ factorizes in FA as $g_1 \cdot g_2$ with $g_2: (X, F) \to (Z, F)$ and $g_1: (Z, F) \to (Y, F)$, we can define on Z topologies K, K' by $K =$ [subsets A of the form $g_1^{r}(s)$ for some s of S], $K' =$ [subsets of the form $g_1^{r}(s')$ for s' from S']. Then we can check that $(Z, K, K'; F)$ is a bitopological F-algebra, that $g_1: (Z, K, K'; F) \to (Y, S, S'; F)$ and $g_2: (X, T, T' F) \to (Z, K, K'; F)$ are morphisms in $BTFA$ such that $f = g_1 \cdot g_2$ in $BTFA$.

For proving that H raises codomains, if for an object $(X, T, T'; F)$ of

BTFA g: $(X, F) \to (Y, F)$ is an F-homomorphism, we assign to Y the coarsest topology M on Y to get an object $(Y, M, M; F)$ of *BTFA* with g: $(X, T, T'; F) \to (Y, M, M; F)$ a morphism of *BTFA*.

To see that H is latticiel, if $[T_i, T_i']$ is a family of bitopologies on X such that each $(X, T_i, T_i'; F)$ belongs to *BTFA*, it is clear that $(X, \Pi(T_i), \Pi(T_i'); F)$ is also in *BTFA*. Thus part (a) of condition (iii), Definition A.1.6, is true. If we are given a family of morphisms f_i: $(X, T, T'; F) \to (Y, S_i, S_i'; F)$, all having the same domain in *BTFA* such that H takes every one of them to the same f: $(X, F) \to (Y, F)$ in *FA*, this f is the $\Pi(f_i)$ from $(X, T, T'; F)$ to $(Y, \Pi(S_i), \Pi(S_i'); F)$, to prove part (b). Finally, a morphism f: $(X, T, T'; F) \to (Y, S, S'; F)$ of *BTFA* belongs to *BTF*A* iff both (X, F) and (Y, F) are F^*-algebras; and this is just the condition for $H(f)$ of *FA* to belong to *F*A*. This completes all requirements to use Lemma A.1.1 for this case.

4. For $G = PO$, consider the functor G^*: *POF*D*A* $\to$ *POFDA*. Using Lemma A.1.1 again, with *F*D*A*, *FDA*, *POF*D*A*, and *POFDA* as A, B, C, and D, respectively, we verify that the functor H: *POFDA* $\to$ *FDA* has the requisite properties to apply the lemma. This verification is quite easy. Since we already know that the inclusion functor from *F*D*A* to *FDA* has a left-adjoint, we deduce that G^* has a left-adjoint by use of the lemma.

5. Since the base functor $G \to S$ is seen to satisfy the properties required of H in Lemma A.1.1, when we take *GFA*, *FA*, G, and S for A, B, C, and D we can deduce that H': *GFA* $\to G$ has a left-adjoint, since the inclusion functor from *FA* to S has a left-adjoint. □

APPENDIX TWO

ALGEBRAIC FUNCTORS

The way we define the algebraic functor, all 25 of the functors mentioned in Theorem A.1.2 would be algebraic functors; essentially, all 25 are functors that forget part or all of an algebraic structure.

Definition A.2.1 A functor $G: A \to B$ is called an *algebraic functor* if (i) G has a left-adjoint $F: B \to A$; (ii) given (H, η, δ) the monad on B defined by the triad (A, F, G), and (A^*, F^*, G^*) the other triad associated with this monad, as in Theorem 7.7.1, there is an isomorphism $L: A \to A^*$ such that $L \cdot F = F^*$ and $G^* \cdot L = G$.

Definition A.2.2 We say that the category A is *an algebra over* the category B (through the functor G) if there is an algebraic functor $G: A \to B$.

For the main theorem of Beck regarding conditions for a functor to be algebraic, we require some basic notions.

Definition A.2.3 Given a functor $G: A \to B$ and a pair of *parallel morphisms* (f, g) of A (that is, morphisms with a common domain and a common codomain), if $(G(f), G(g))$ have a coequalizer (e, C) in B, G is said to *raise the coequalizer uniquely* if there is a unique pair (e^*, C^*) in A with $G(e^*) = e$ and $G(\mathrm{C}^*) = \mathrm{C}$, and further this (e^*, C^*) is a coequalizer of (f, g).

Definition A.2.4 A coequalizer (e, C) of a parallel pair (f, g) of A is called an *absolute coequalizer* if, for any functor $F: A \to B$, $(F(e), F(C))$ is a coequalizer of $(F(f), F(g))$ in B.

Definition A.2.5 A triple $(f, g; e)$ consisting of a parallel pair of morphisms f, g of A and a morphism e such that $e \cdot f = e \cdot g$ is called a *fork* in A; the codomain of e is called the codomain of the fork.

Definition A.2.6 A fork $(f, g; e)$ in A is called a *split fork* with splitting morphisms (s, t) if there are morphisms s, t such that $e \cdot s = 1_{\mathrm{dom}(s)}$, $f \cdot t = 1_{\mathrm{dom}(t)}$, and $g \cdot t = s \cdot e$.

It is easy to see that for a split fork $(f, g; e)$ with codomain C, (e, C) is a coequalizer of (f, g); it is even an absolute coequalizer, since the image by a functor of a split fork is surely a split fork. We call (e, C) the split coequalizer of (f, g) in this case. We now state *Beck's theorem*:

Theorem A.2.1 *Either one of the following (equivalent) conditions is necessary and sufficient for a functor $G: A \to B$ to be algebraic when G has a left-adjoint:*

(a) *for any parallel pair (f, g) of A, when $(G(f), G(g))$ has an absolute coequalizer in B, G raises it uniquely;*
(b) *for any parallel pair (f, g) of A, when $(G(f), G(g))$ has a split coequalizer in B, then G raises it uniquely.*

PROOF. (i) *If G is algebraic, then* (a) *is true*. Since A, A^* are isomorphic, it would suffice to check (a) for A^* and $G^*: A^* \to B$. Let (A, a), (B, b) be objects of A^*, and f, g be morphisms from $\hom_{A^*}((A, a), (B, b))$, such that the pair $(G^*(f), G^*(g))$ or (f, g) in B has an absolute coequalizer (e, C): then $(H(f), H(g))$ also has as coequalizer $(H(e), H(C))$. Since f, g are morphisms of A^*, we have the relations $b \cdot H(f) = f \cdot a$ and $b \cdot H(g) = g \cdot a$; and because e, $H(e)$ are coequalizers we have $e \cdot f = e \cdot g$ and $H(e) \cdot H(f) = H(e) \cdot H(g)$. Hence, $(e \cdot b) \cdot H(f) = e \cdot f \cdot a = e \cdot g \cdot a = (e, \cdot b) \cdot H(g)$; but then, from the definition of $H(e)$ as a coequalizer of $H(f)$, $H(g)$, there must be a unique morphism $c: H(C) \to C$ in B such that $c \cdot H(e) = e \cdot b$. (See Figure A.2.) If (C, c) is an object of A^*, it would follow that e is a morphism in A^* from (B, b) to (C, c). So we now check that (C, c) is an object of A^*. (See Figure A.3.) For this we have to show that the outer square and outer triangle in the figure give commutative diagrams; we know the inner square and inner triangle are commutative, since (B, b) is in A^*. From the definition of c in terms of b, and the naturalness of η and δ, the four trapezia of the top part and the three trapezia of the bottom part give commutative diagrams. Hence from the top part we get $c \cdot H(c) \cdot H^2(e) = c \cdot H(e) \cdot H(b) = e \cdot b \cdot H(b) =$

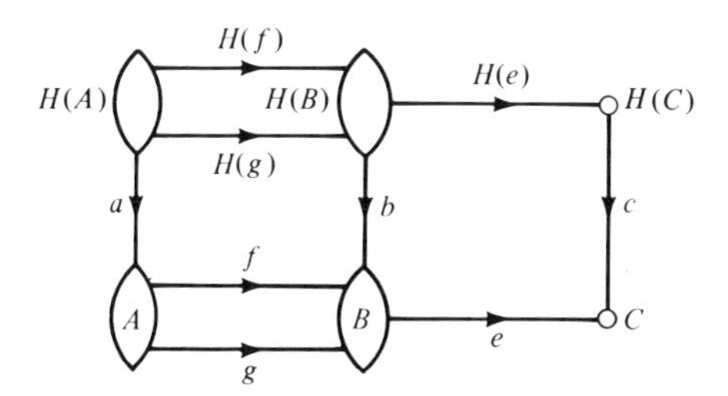

Figure A.2

$e \cdot b \cdot \delta(B) = c \cdot H(e) \cdot \delta(B) = c \cdot \delta(C) \cdot H^2(e)$; as $H^2(e)$ is a coequalizer of $(H^2(f), H^2(g))$, it is a regular epimorphism and so right-cancellative; hence we get $c \cdot H(c) = c \cdot \delta(C)$. Similarly, from the bottom part we get $c \cdot \eta(C) \cdot e = c \cdot H(e) \cdot \eta(B) = e \cdot b \cdot \eta(B) = e \cdot 1_B = 1_C \cdot e$. Again e is a regular epimorphism, therefore right-cancelable; thus $c \cdot \eta(C) = 1_C$. These prove that (C, c) is in A^*. So we have only to check that $(e, (C, c))$ is a coequalizer of (f, g) in A^* to prove (a): we already have $e \cdot f = e \cdot g$. Suppose $k \cdot f = k \cdot g$ for some k: $(B, b) \to (D, d)$ in A^*. Since k is a morphism in A^*, we have $k \cdot b = d \cdot H(k)$; we already have $e \cdot b = c \cdot H(e)$. Because (e, C) is a coequalizer of (f, g) in B, $k \cdot f = k \cdot g$ implies that there is a unique k': $C \to D$ in B such that $k = k' \cdot e$; but then, since

Figure A.3

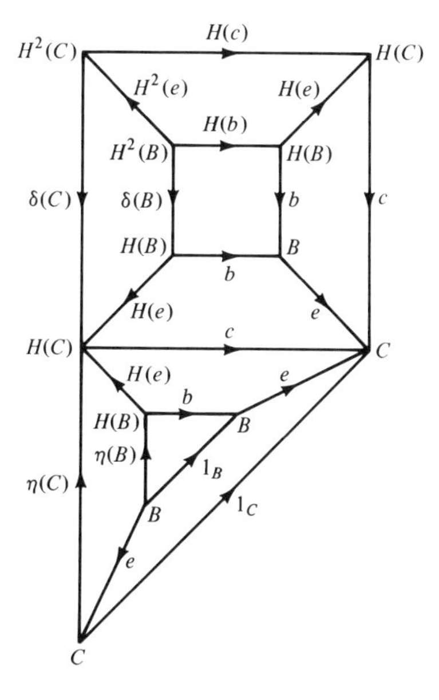

H is a functor, it follows that $H(k) = H(k') \cdot H(e)$. Hence we have $k' \cdot c \cdot H(e) = k' \cdot e \cdot b = k \cdot b = d \cdot H(k) = d \cdot H(k') \cdot H(e)$. Since (e, C) is an absolute coequalizer, $(H(e), H(\mathrm{C}))$ is also an equalizer, so that $H(e)$ is a regular epimorphism and is right-cancelable. Hence we get finally $k' \cdot c = d \cdot H(k')$, showing that k' is a morphism of A^* from (C, c) to (D, c) such that $k' \cdot e = k$. The uniqueness of this k' follows from the uniqueness of the k' in B. Hence $(e, (\mathrm{C}, c))$ is a coequalizer of (f, g) in A^*, and is evidently the unique pair from A^* and a coequalizer for which $G^*(e) = e$ in B and $G^*(\mathrm{C}, c) = \mathrm{C}$ in B. Thus (a) has been proved for G^*.

(ii) *Condition* (a) *implies condition* (b). This is easy to see since a split coequalizer is also an absolute coequalizer.

(iii) *Condition* (b) *implies that* $G: A \to B$ *is an algebraic functor.* For proving this, let us first create a category $\mathrm{Tr}(H, \eta, \delta)$ from a monad (H, η, δ) on B, with a typical object being any triad (A', F', G') determining the monad [through a near-equivalence $(\eta, \nu'; F', G')$] and a typical morphism $L: (A', F', G') \to (A, F, G)$ as a functor $L: A' \to A$ such that $L \cdot F' = F$ and $G \cdot L = G'$. [Then by 6(a) under Exercises and Remarks on Chapter 7, we also have $L\nu' = \nu L$.] We have already seen that in (A^*, F^*, G^*) we have a final object; so if there were another final object it would be isomorphic with this; hence, we merely show that when G: $A \to B$ satisfies condition (b), (A, F, G) is also a final object in this category, whence it would follow that A, A^* are isomorphic or G is algebraic.

Let (A', F', G') be any triad from $\mathrm{Tr}(H, \eta, \delta)$ and (A, F, G) one satisfying condition (b); let $(\eta, \nu; f, g)$ be the near-equivalence through which (A, F, G) determines the monad (H, η, δ). We seek a morphism M that is, moreover, unique from (A', F', G') to (A, F, G). If M is a morphism from (A', F', G') to (A, F, G), then we know that $M\nu' = \nu M$, $MF'G' = FG' = FGM$. (See Figure A.4.)

Any object A' of A' determines a fork $[\nu'F'G'(\mathrm{A}'), F'G'\nu'(\mathrm{A}'); \nu'(\mathrm{A}')]$ in A', and applying M to this would give the similar fork determined by $\mathrm{A} = M(\mathrm{A}')$. Further, the transform of the first fork by G' must be the same as the transform of the second fork by G, since $G = M \cdot G'$. These both give a fork in B that is a split fork, namely, $[G\nu FG'(\mathrm{A}'), HG'\nu'(\mathrm{A}'); G'\nu'(\mathrm{A}')]$ with the splitting morphisms $[\eta G'(\mathrm{A}'), \eta HG'(\mathrm{A}')]$; for we saw in the proof of Theorem 7.7.1 that the morphism L: $(A', F', G') \to (A^*, F^*, G^*)$ takes A' to $(G'(\mathrm{A}'), \nu'G'(\mathrm{A}'))$ of A^*, and it is easy to see that for any (A, a) of $O(A^*)$ we have a split fork $[\delta(\mathrm{A}), Ha, a]$ with splitting morphisms $[\eta(\mathrm{A}), \eta H(\mathrm{A})]$. So by condition (b) for G there must be a unique pair (k, C) in A with $G(\mathrm{C}) = G'(\mathrm{A}')$ and $G(k) = \nu'G'(\mathrm{A}')$, and this (k, C) should also be a coequalizer for $[\nu FG'(\mathrm{A}'), FG'\nu'(\mathrm{A}')]$; it follows that this C must be $M(\mathrm{A}')$ and k must be $\nu M(\mathrm{A}') = M\nu'(\mathrm{A}')$. These prove that if M is defined, it is unique! They also show that if defined, M must

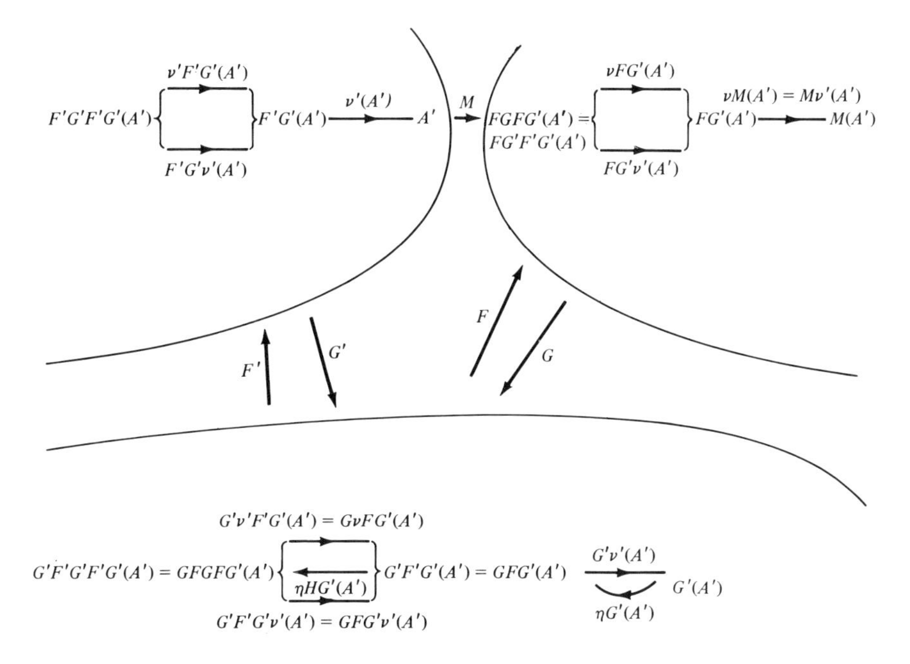

Figure A.4

satisfy $M(\mathrm{A}') = \mathrm{C}$ for the unique (k, C) that raises the coequalizer of the split fork in B as above.

So it remains to define the M suitably for morphisms in A'. For f': $\mathrm{A}' \to \mathrm{B}'$ in A', we have the commutativity of the two left squares in Figure A.5, and $k(\mathrm{A}')$, $k(\mathrm{B}')$ are coequalizers. It follows that $k(\mathrm{B}') \cdot FG'(f')$ must factorize uniquely as $M(f') \cdot k(\mathrm{A}')$ for a unique $M(f')$ from $M(\mathrm{A}')$ to $M(\mathrm{B}')$. This gives M for f' and is easily seen to make M a functor from (A', F', G') to (A, F, G). Since its uniqueness has already been noted, (A, F, G) is a final object in $\mathrm{Tr}(H, \eta, \delta)$. So there is an isomorphism of (A, F, G) on (A^*, F^*, G^*); thus G is an algebraic functor. □

We shall use the Theorem A.2.1 to prove that a great many functors are algebraic.

Lemma A.2.1 *All the 25 functors mentioned in Theorem A.1.2, part (i), are algebraic functors.*

PROOF. As a typical case let us consider the functor G^*: $GF^*D^*A \to GFDA$, where G denotes any one of the types CSU, SU, BT, PO, or S, and F^*, D^* are assumed to contain F, D, respectively. Thus G^* is an inclusion functor. From Theorem A.1.2 we already have a left-adjoint for

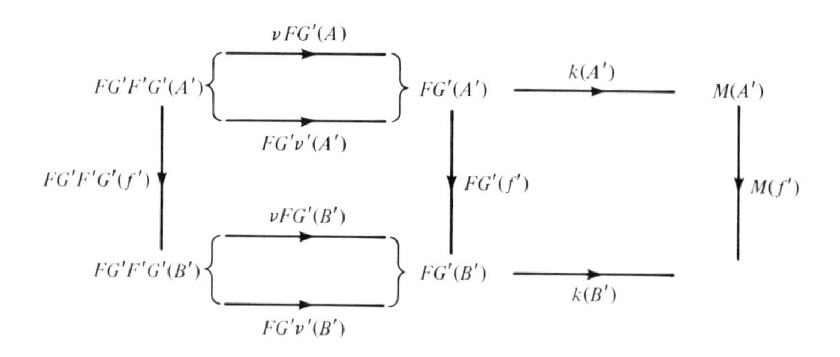

Figure A.5

G^*. Hence we need only verify condition (b) of Theorem A.2.1 for this G^*.

Let then (f, g) be a pair of parallel morphisms in GF^*D^*A between the objects (X, G, F^*) and (Y, H, F^*), both GF^*-algebras, and suppose the pair (f, g) considered as $(G^*(f), G^*(g))$ in $GFDA$ have a split coequalizer $(e, (Z, K, F))$ in $GFDA$. Given a typical operation $(u, 0)$ from F^*_0, that is, a nullary operation from F^*, we set $u^Z(\varnothing) = e(u^Y(\varnothing))$. For a typical operation (θ, n, v) from F^*_1 we know there is a functor $P(\theta) = P[(1)^{v(1)}, \ldots, (1)^{v(n)}]$ of $GFDA$ in itself that takes a typical object (X, G, F) to $P[(X, G)^{v(1)}, \ldots, (X, G)^{v(n)}]$ and a typical morphism f: $(X, G, F) \to (Y, H, F)$ to $P[f^{v(1)}, \ldots, f^{v(n)}]$. Since $(e, (Z, K, F))$ is a split coequalizer, it is also an absolute coequalizer; hence we have $[P(\theta)(e), P(\theta)(Z, K, F)]$ is a coequalizer of $[P(\theta)(f), P(\theta)(g)]$ in $GFDA$. Also, θ^X and θ^Y are morphisms from $P(\theta)(X, G, F)$ to (X, G, F) and from $P(\theta)(Y, H, F)$ to (Y, H, F); that f, g are morphisms in $GFDA$ means that we have the commutativity relations $\theta^Y \cdot P(\theta)(f) = f \cdot \theta^X$ and $\theta^Y \cdot P(\theta)(g) = g \cdot \theta^X$. Hence we have $e \cdot \theta^Y \cdot P(\theta)(f) = e \cdot f \cdot \theta^X = e \cdot g \cdot \theta^X$ [since e is a coequalizer of (f, g)] $= e \cdot \theta^Y \cdot P(\theta)(g)$. But we have a coequalizer $[P(\theta)(e), P(\theta)(Z, K, F)]$ for $(P(\theta)(f), P(\theta)(g))$; hence there must be a unique morphism, which we define to be θ^Z from $P(\theta)(Z, K, F)$ to (Z, K, F) such that $e \cdot \theta^Y = \theta^Z \cdot P(\theta)(e)$. Then e is clearly a θ-homomorphism of the algebra (Y, H, F) in (Z, K, F). Similarly we can extend all the operations from F^* to Z in such a way that we get a GF^*-algebra (Z, K, F^*) with e an F^*-homomorphism from (Y, F^*) to (Z, F^*). Since we assumed that $(e, (Z, \ldots))$ was a split coequalizer, e must be a surjective homomorphism of (Y, F) on (Z, F); since it also preserves the operations from F^*, and since (Y, F) is an F^*D^*-algebra, it follows that (Z, F^*) should also be an F^*D^*-algebra. Thus e is a morphism in GF^*D^*A between the objects (Y, H, F^*) and (Z, K, F^*) from that category. Since G^* is an inclusion functor, the fact that e is a coequalizer of f, g in $GFDA$ would now show that $(e, (Z, K, F^*))$ is a coequalizer of f, g in GF^*D^*A. Evidently $(e, (Z, K, F^*))$ is the unique pair from GF^*D^*A that goes by G^* to $(e, (Z, K, F))$. Thus we have proved condition (b), and the theorem, for the particular case. (Note that there is essentially only one way in which the operations of F^* can be extended to Z, since they are already defined in Y and we have the relation $e \cdot \theta^Y = \theta^Z \cdot F(\theta)(e)$, and $F(\theta)(e)$ is an epimorphism.) Other cases of the theorem can be similarly treated. □

APPENDIX THREE

TOPOLOGICAL FUNCTORS

Topological functors are defined essentially as in the book by Herrlich and Strecker [15].

Definition A.3.1 A *source* $(X, f_i\colon i \text{ in } I)$ in the category C consists of an object X and an indexed family (f_i) of morphisms with domain X. When (Y, f) is another source with a single morphism f, and X is the codomain of f, we have a composite source $(Y, f_i \cdot f\colon i \text{ in } I) = (X, f_i\colon i \text{ in } I) \cdot (Y, f)$.

Definition A.3.2 The category C is called an $(E, M\#)$-category if (i) E is a class of epimorphisms of C closed for compositions with isomorphisms and $M\#$ is a class of sources in C also closed for compositions with isomorphisms; (ii) C is $(E, M\#)$-factorable—that is, any source of C can be factored as $(X, f_i) \cdot (Y, f)$ with f in E and (X, f_i) in M; and (iii) C has the $(E, M\#)$-diagonalization property—that is, when a source has two factorizations $(Z, f_i) \cdot (X, e) = (Y, m_i) \cdot (X, f)$, with e in E and (Y, m_i) in $M\#$, there exists a (unique) morphism $g\colon Z \to Y$ such that $g \cdot e = f$ and $m_i \cdot g = f_i$ for each i.

The diagonalization property would ensure uniqueness of an $(E, M\#)$-factorization of a source up to a middle isomorphism (g).

Definition A.3.3 Given an $(E, M\#)$-category C and a functor $F\colon A \to C$, we say that:

(a) a source $(\mathrm{A}, f_i\colon \mathrm{A} \to \mathrm{A}_i)$ in A is an *F-initial source* iff for each source $(\mathrm{A}', g_i'\colon \mathrm{A}' \to \mathrm{A}_i)$ in A and each morphism $f\colon F(\mathrm{A}') \to F(\mathrm{A})$

in C for which $F(f_i) \cdot f = F(g_i')$ for each i, there exists a unique morphism f^*: A′ → A in A such that $F(f^*) = f$ and $f_i \cdot f^* = g_i'$ for each i;

(b) a source $(A, f_i: A \to A_i)$ in A *F-lifts a source* $(X, g_i: X \to F(A_i))$ in C iff there exists an isomorphism $h: X \to F(A)$ in C such that $F(f_i) \cdot h = g_i$ for each i;

(c) F is an *(E, M#)-topological functor* (from A to C) iff for each family $[A_i$: i in $I]$ of objects from A and each source $(X, g_i$: $X \to F(A_i))$ in $M\#$, there exists an F-intial source $(A, f_i: A \to A_i)$ in A that F-lifts (X, g_i).

After these basic definitions from Herrlich's paper, we establish a general result that enables us to recognize many of the functors in Chapter 7 to be topological ones [for a suitable $(E, M\#)$].

Definition A.3.4 A triple (A, E, M) is called a *top category* if A is a (small) complete category that admits unique (E, M)-factorization, as defined in Section 6.3.

Definition A.3.5 A functor $F: A \to B$ is called a *top functor* from (A, E, M) to (B, E^*, M^*) if (i) F preserves all products, (ii) [A in $O(A)$, m^* in $\hom_B(-, F(A))$] ⇒ [there is an m in $\hom_A(-, A)$ with $F(m) = m^*$, and also f in $\hom_A(-, A)$ and $F(f) = m^* \cdot h^*$ imply that $F(h) = h^*$ and $f = m \cdot h$ for a suitable h of $M(A)$].

Theorem A.3.1 *(a) A top category (A, E, M) is an (E, M#)-category in Herrlich's sense, where M# consists of sources of the form* $[A, p_i \cdot m]$ *where* p_i: $P(A_i) \to A_i$ *denotes the canonical morphism of a product* $P(A_i)$ *of the codomains of the* $(p_i \cdot m)$ *on the ith factor, and m is a morphism from M in* $\hom_A(A, P(A_i))$.

(b) A top functor F: (A, E, M) → (B, E,M*) is an (E*, M*#)-topological functor in the sense of Herrlich.*

PROOF. (a) Given a source $(A, f_i: A \to A_i)$ in A, and a product $P(A_i)$ of the family (A_i) with the p_i: $P(A_i) \to A_i$, we have a first resolution of the given source in the form $[P(A_i), p_i] \cdot [A, P(f_i)]$. Factorizing the morphism $P(f_i)$: $A \to P(A_i)$ in the form $P(f_i) = m \cdot e$ (with m, e from M, E), and A′ as the domain of m (and codomain of e), we get another resolution of the source as $[A', p_i \cdot m] \cdot [A, e]$; this is an $(E, M\#)$-factorization of the given source. To prove the diagonalization property, if we have two equal factorizations of a source in the form $[Z, f_i] \cdot [X,e] = [Y, p_i \cdot m] \cdot [X, f]$ with e in E and $[Y, p_i \cdot m]$ from $M\#$, so that m: $Y \to P(A_i)$, p_i: $P(A_i) \to A_i$, $f: X \to Y$, $e: X \to Z$, and f_i: $Z \to A_i$, where $P(A_i)$ is a product in A of the object family $[A_i]$, the source $[Z, f_i]$ can be factored as

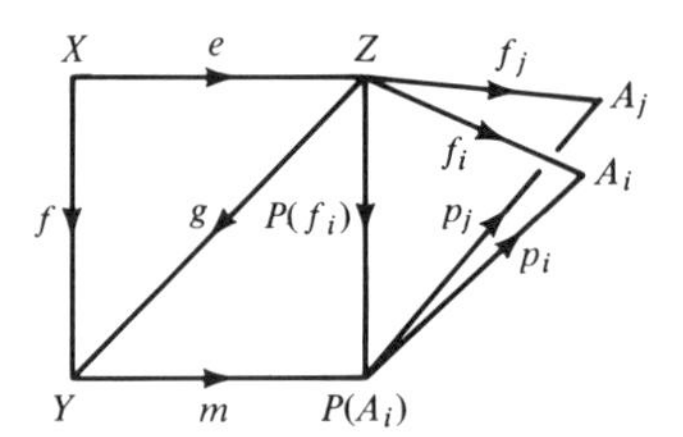

Figure A.6

$[P(A_i), p_i] \cdot [Z, P(f_i)]$ (see Figure A.6). From the first we have $p_i \cdot m \cdot f = f_i \cdot e$ for each i; then using the second we have $f_i \cdot e = p_i \cdot P(f_i) \cdot e$; hence $p_i \cdot m \cdot f = p_i \cdot P(f_i) \cdot e$ for each i. Then the definition of the product $[P(A_i), p_i]$ implies that $m \cdot f = P(f_i) \cdot e$. Our assumption regarding (E, M) in A then gives a unique $g: Z \to Y$ such that $g \cdot e = f$ and $m \cdot g = P(f_i)$; the latter gives $(p_i \cdot m) \cdot g = p_i \cdot P(f_i) = f_i$ for all i. Hence we get the $(E, M\#)$-diagonalization property for sources. Thus A is an $(E, M\#)$-category.

(b) Given the top functor $F: (A, E, M) \to (B, E^*, M^*)$ and a family of objects (A_i) of A, we form a product source $[P(A_i), p_i]$ in A; F takes this to a product source $[F(P(A_i)) = aP(F(A_i)), F(p_i)]$. If we are now given a source from M^* of the form $[B, P^*(i) \cdot m^*: B \to F(A_i)]$ in B where m^*

Figure A.7

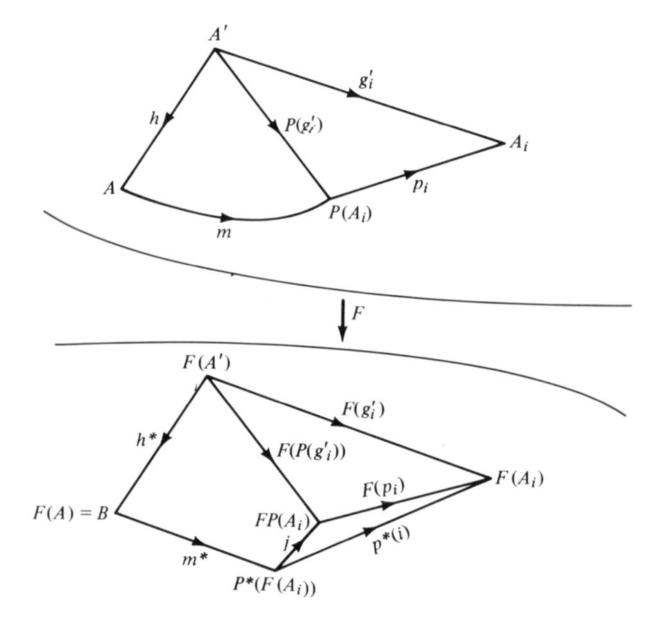

is a morphism of M^* from B to a product object $P^*[F(\mathrm{A}_i)]$ and $[P^*(F(\mathrm{A}_i), p^*(i)]$ is the source given by that product of the $[F(\mathrm{A}_i)]$ in B, then since we have two products in B for the $[F(\mathrm{A}_i)]$, there is an isomorphism j: $P^*[F(\mathrm{A}_i)] \to F(P(\mathrm{A}_i))$ such that $F(p_i) \cdot j = p^*(i)$ for each i. (See Figure A.7.) Now $j \cdot m^*$: $\mathrm{B} \to F(P(\mathrm{A}_i))$ is from M^*, hence by our assumption that F is a top functor there is an m: $\mathrm{A} \to P(\mathrm{A}_i)$ in A with the properties given in condition (ii) of the definition of the top functor (Definition A.3.5). Evidently the source $[\mathrm{A}, p_i \cdot m]$ F-lifts the source $[\mathrm{B}, F(p_i) \cdot j \cdot m^*] = [\mathrm{B}, p^*(i) \cdot m^*]$; so it remains only to check that this source $[A, p_i \cdot m]$ is an F-initial source. Given then a source $[\mathrm{A}', g_i'$: $\mathrm{A}' \to \mathrm{A}_i]$ in A and a morphism h^*: $F(\mathrm{A}') \to F(\mathrm{A}) = \mathrm{B}$ such that $p^*_i \cdot m^* \cdot h^* = F(g_i')$ for each i, since there is a factorization of the source $[A', g_i']$ in the form $[P(\mathrm{A}_i), p_i] \cdot [\mathrm{A}', P(g_i')]$, which goes down under F to a factorization $[F(P(\mathrm{A}_i)), F(p_i)] \cdot [F(\mathrm{A}'), F(P(g_i'))]$, the property for m relative to m^* ensures that there is an h: $\mathrm{A}' \to \mathrm{A}$ such that $F(h) = h^*$ and $m \cdot h = P(g_i')$. This means that $[\mathrm{A}', g_i']$ factors as $[\mathrm{A}, p_i \cdot m] \cdot [\mathrm{A}', h]$, with $F(h) = h^*$. This proves that the source $[\mathrm{A}, p_i \cdot m]$ is F-initial, and completes the proof that F is $(E^*, M^*\#)$-topological. □

We now apply the last result to check that many of the functors we had before are topological for suitable (E, M)'s.

Lemma A.3.1 *(a) For suitable choice of (E, M) in each case, the categories CSUK, SUK, BTK, POK, and K, where K stands for any one of FDA, FA, or S, are top categories.*

(b) Further, keeping the K part fixed, there are functors C: CSUK → SUK, B: SUK → BTK, P: BTK → POK, and S: POK → K that are all top functors, and so topological functors [for suitable (E, M)'s in the categories involved].

PROOF. (a) We shall sketch the proof for the case of $K = FDA$; the other two choices of K are easier, and similar.

$(CSUFDA, E(U), M(U))$ is a top category when we take $E(U)$ to consist of morphisms f: $(X, U, F) \to (Y, V, F)$ for which f is a uniform map of (X, U) on a subset of (Y, V) that is dense in (Y, V) under the topology $T[V \vee V^r]$, and of those morphisms that differ from such by an extra isomorphism factor appearing on the left or the right; while $M(U)$ consists of morphisms that differ by at most an isomorphism factor, as just stated, from those of the form m: $(X, U, F) \to (Y, V, F)$, where X is a subset of Y, m is the inclusion morphism and an F-homomorphism [so that (X, F) is a subalgebra of (Y, F)], U is the carry-back or relative semiuniformity from V, and $(X, T(U \vee U^r))$ is a closed subspace of $(Y, T(V \vee V^r))$.

For $SUFDA$ we have two top category structures to consider: $E(U)$, $M(U)$, similar to the definition in the preceding paragraph; and $E^*(U)$,

$M^*(U)$, which consist respectively of surjective and injective uniform F-homomorphisms, with the latter using a carry-back semiuniformity as in $M(U)$, and of those that differ from these by an isomorphism factor.

($BTFDA$, $E(B)$, $M(B)$) is a top category where $E(B)$ consists of morphisms of $BTFDA$ that are surjective as set maps, while $M(B)$ consists of those that are injective as set maps; moreover (for $M(B)$), the two topologies on the domain are the carry-back ones (using the set map) from those of the codomain.

($POFDA$, $E(P)$, $M(P)$) is a top category where $E(P)$, $M(P)$ are defined similarly to $E(B)$, $M(B)$ in the preceding paragraph; in the case of morphisms in $M(P)$, the preorder of the domain should be the carry-back from that of the codomain.

(FDA, E, M) is a top category when E consists of surjective F-homomorphisms and M consists of injective F-homomorphisms.

We leave the verifications that these indeed give top categories to the reader.

(b) Then we claim that the following are top functors: C: $(CSUFDA, E(U), M(U)) \to (SUFDA, E(U), M(U))$; B: $(SUFDA, E^*(U), M^*(U)) \to (BTFDA, E(B), M(B))$; P: $(BTFDA, E(B), M(B)) \to (POFDA, E(P), M(P))$; S: $(POFDA, E(P), M(P)) \to (FDA, E, M)$.

Since all these functors have left-adjoints (by Theorem A.1.1), they preserve all limits, hence all products. In each case the proof of property (ii) involves showing that given sets X, Y with X a subset of Y, if the functor is an F: $(GFDA, E(G), M(G)) \to (HFDA, E(H), M(H))$, when (X, G_1) is obtained from (Y, G_2) by a carry-back process using the inclusion map, then their images under F give an (X, H_i) obtained from the (Y, H_2) again by a carry-back using the inclusion map of X in Y. For instance, for C if (X, U, F) is a $T(V \vee V^r)$-closed subset of (Y, V, F) and U is the relative semiuniformity on X defined by V, then for the associated bitopologies (T, T') for X and (T^*, T'^*) for Y, (X, T, T') is a $(T^* \vee T'^*)$-closed subset of Y and T, T' are the relative topologies for X determined by T^*, T'^* on Y. Similarly, for P if (X, T, T'), (Y, T^*, T'^*) are such that X is a subset of Y with the relative topologies T, T' on X defined by T^*, T'^*, and X is a subalgebra of the FD-algebra Y, then the preorder $\leq(T)$ on X is the relative preorder on X-determined by the preorder $\leq(T^*)$ on Y. The details of the proofs are left for the reader to complete. □

BIBLIOGRAPHY

1. M. Barr and J. Beck, Acyclic models and triples, *Proc. Conf. Categ. Alg.* (La Jolla, 1965), pp. 336–344. New York: Springer, 1966.
2. G. Birkhoff, *Lattice Theory,* 3d ed. Providence, RI: Amer. Math. Soc. Publ. 25, 1967.
3. R. H. Bruck, A survey of binary systems, *Ergeb. Math. u Grenz.* 20 (1958).
4. H. Cartan and S. Eilenberg, *Homological Algebra.* Princeton, NJ: Princeton Univ. Press, 1956.
5. A.H. Clifford and G. B. Preston, *Algebraic Theory of Semigroups,* Vols. I and II. (Amer. Math. Soc. Survey 7, 1967).
6. P. M. Cohn, *Universal Algebra.* New York: Harper and Row, 1965.
7. P. Dubreil, *Algèbre,* Vol. 1. Paris: Gauthier Villars, 1946.
8. S. Eilenberg and S. MacLane, General theory of natural equivalences, *Trans. Amer. Math. Soc.* 58, 231–294, 1945.
9. S. Eilenberg and J. C. Moore, Adjoint functors and triples, *Ill. J. Math.* 9, 381–398, 1965.
10. S. Eilenberg and N. E. Steenrod, *Foundations of Algebraic Topology.* Princeton, NJ: Princeton Univ. Press, 1952.
11. P. Freyd, *Abelian Categories: An Introduction to the Theory of Functors.* New York: Harper and Row, 1964.
12. G. Grätzer, *Universal Algebra.* Princeton, NJ: Van Nostrand, 1968.
13. H. Herrlich, Topologische Reflexionen und Coreflexionen, *Lect. Notes Math.* 78, 1968.
14. H. Herrlich, Topological functors, *Gen. Top. and Appl.* 4, 125–142, 1974.
15. H. Herrlich and G. E. Strecker, *Category Theory.* Boston: Allyn and Bacon, 1973.

16. P. C. Kainen, Weak adjoint functors, *Math. Z.* 122, 1–9, 1971.
17. D. M. Kan, Adjoint functors, *Trans. Amer. Math. Soc.* 87, 294–329, 1958.
18. H. Kleisli, Every standard construction is induced by a pair of adjoint functors, *Proc. Amer. Math. Soc.* 16, 544–546, 1965.
19. V. S. Krishnan, Additive asymmetric semiuniform spaces and semigroups, *J. Madras Univ.* B, 175–198, 1961,
20. V. S. Krishnan, Semiuniform spaces and seminorms, semimetrics and semiecarts in *APO*-semigroups, *Gen. Top. and Appl. Mod. Alg. and Anal.* (Proc. Kanpur Conf.), pp. 163–171, 1968.
21. E. S. Ljapin, *Semigroups.* Prividence, RI: Amer. Math. Soc. Translat. Math. Monogr. 3, 1963.
22. S. MacLane, *Categories for the Working Mathematician* (Graduate Texts in Math. 5). New York: Springer, 1971.
23. H. M. MacNeille, Partially ordered sets, *Trans. Amer. Math. Soc.* 42, 1937.
24. A. Malčev, Uber die Einbettung von assoziativen Systemen in Gruppen, I, II, *Rec. Math.* (N.S.) 6, 1948; 8, 1950.
25. B. Mitchell, *Theory of Categories.* New York: Academic Press, 1965.
26. L. Nachbin, *Topology and Order* (Van Nostrand Studies 4). Princeton, NJ: Van Nostrand, 1965.
27. B. Pareigis, *Categories and Functors.* New York: Academic Press, 1970.
28. D. Tamari, Monoides préordonnés et chaines de Malčev, *Bull. Soc. Math. Fr.* 82, 1954.
29. A. Tychonoff, Uber die topologische Erweiterung von Raümen, *Math. Ann.* 102, 1930.
30. R. Vaidyanathaswamy, *Treatise on Set Topology.* Madras, 1947; 2nd ed. New York: Chelsea, 1970.
31. A. Weil, Sur les espaces à structure uniforme et sur la topologie générale, *Act. Sci. Ind.* 55, 1937.

INDEX